COMMODITIES IN THE
INTERNATIONAL ECONOMY

Series Editors: Bill Albert and Adrian Graves

Tin in the World Economy

JOHN T. THOBURN

EDINBURGH UNIVERSITY PRESS

To the memory of my father,
Thomas Thoburn

© John T. Thoburn, 1994

Edinburgh University Press Ltd
22 George Square, Edinburgh

Typeset in Linotron Ehrhardt
by Koinonia Ltd, Bury, and
printed and bound in Great Britain
by The University Press, Cambridge

A CIP record for this book is available
from the British Library

ISBN 0 7486 0516 9

Contents

Preface

When Bill Albert, co-editor of this series, invited me to contribute a volume on tin, I had not worked on the industry for a decade and was involved in research on manufacturing in China and Japan. Yet, an opportunity to rediscover the tin industry was exciting. During that decade away, the International Tin Agreement had collapsed in a spectacular default. Brazil had arisen as an important new producer, and the old-established producers in South East Asia were restructuring in the hope of survival in the face of historically low tin prices. Bolivia, the other major tin exporter, was reorganising its tin industry as part of a structural adjustment programme for the whole of its long-ailing economy. Writing the book has involved field trips to Thailand, Malaysia and Indonesia, in February and March 1992, and to Brazil in September 1992; and I am extremely grateful to the Nuffield Foundation for a grant which made those visits possible. I should also like to thank the University of East Anglia for finance for some interviewing in the UK and for a visit to UNCTAD in Geneva.

During my visit to Thailand, I received help and hospitality from Chee Peng Lim of ESCAP in Bangkok, and Amy Chee, for which I am most grateful; I should also like to thank Amy for her patience in letting me use my stay to improve my Chinese! I appreciate too the hospitality received from Ishak Omar of the Universiti Pertanian Malaysia and his family, from K. W. Thee in Jakarta, and Marcos and Lucia Borges in Rio. In Malaysia, Jahaya Mat and his staff at the Ministry of Primary Industries provided much cheerful, practical help. In Brazil, Diana Kinch, *Metal Bulletin*'s Latin American correspondent, did much to make my visit useful. Roderick MacLeod kindly shared with me his long experience of the tin industry in Asia and Latin America, and took me with him on a field trip to Rondonia in the Amazon. In London, *Metal Bulletin*'s non-ferrous metals editor,

Ann-Marie Moreno, generously helped with contacts in the industry, as did Robin Amlôt, editor of the tin industry's excellent trade magazine, *Tin International*. I thank all the miners and officials who gave me information, including officials of the Department of Mineral Resources in Bangkok, ESCAP's Mineral Resources Section, the Mines Department in Kuala Lumpur, and the National Department of Mineral Production in Brasilia. I especially thank the Indonesian state tin company, P. T. Timah, its president-director Kuntoro Mangkusubroto, and the London managing director of Indometal, P. Djienawie, for their help in arranging my visit to the tin mining island of Bangka.

I acknowledge with thanks the kind permission of the editor of the Avebury series of Ashgate Publishing (formerly Gower Publishing), the publishers of my earlier book *Multinationals, Mining and Development: A Study of the Tin Industry* (1981), to reproduce some parts of that book in the present volume, usually in shortened and revised form. These mainly appear in chapter 5. I also thank Redzwan Sumun, Executive Secretary of the Association of Tin Producing Countries, for permission to print his poem on the International Tin Agreement, written in the heat of the nego-tiations after the ITA's collapse in 1985. This is included as an addendum to Chapter 6, and gives a flavour of feelings in the industry at that watershed in its development. Some material from the present book was first pub-lished in *Resources Policy*, vol. 20, no. 2, June 1994, pp. 125–33, published by Butterworth-Heinemann, Oxford, UK.

Bill Albert has been a helpful and efficient editor, and I thank him for his comments on a draft of the book. Of course, neither he nor any person or organisation mentioned here necessarily agrees with the views expressed in this book, and any errors are my own. I am grateful for secretarial assis-tance to Sue Rowell, Judy Sparks and Jill Wyatt. Finally, and as ever, June, my wife, has managed to be supportive, despite having to fit it into her own busy schedule of teaching, writing and consultancy.

List of Tables, Figures and Plates

FIGURES

PLATES*

1.   Malaysian gravel pump tin mine
2.   Malaysian gravel pump tin mine (close up)
3.   The palong on a Malaysian gravel pump mine
4.   Tin concentrates delivered to the tin shed for further
      processing, Malaysian gravel pump mine.
5.   Tin ingots, Mentok smelter, Indonesia
6   Tin dredge, Malaysia, depositing tailings behind it as it digs
7.   Offshore tin dredge, Bangka island, Indonesia
8.   Bucket ladder raised for maintenance, offshore tin dredge,
      Indonesia
9.   Dulang washers mining for tin in Malaysia
10.   Suction boat mining tin illegally off coast of southern
      Thailand in 1970s
11.   Mining for tin with excavator, Rondonia, Brazil
12.   Jig plant for primary treatment of tin ore, Rondonia, Brazil

All photographs are the author's.
*Between pp. 18–19.

Abbreviations

AMC	Amalgamated Metal Corporation
AR	annual report
ASEAN	Association of South East Asian Nations
ATPC	Association of Tin Producing Countries
BSM	buffer stock manager (of ITC)
BSMI	Bulletin of Statistics relating to the Mining Industry, Malaysia
CTS	Consolidated Tin Smelters
DLA	Defence Logistics Agency, USA
DME	developed market economy
DMR	Department of Mineral Resources, Thailand
DNPM	Departamento Nacional da Producao Mineral, Brazil
EIU	The Economist Intelligence Unit, London
ESCAP	(United Nations) Economic and Social Commission for Asia and the Pacific, Bangkok
FEER	*Far Eastern Economic Review*, weekly, Hong Kong
FMS	Federated Malay States
GSA	General Services Administration, USA
HoC	House of Commons, London
ICA	international commodity agreement
IFS	International Financial Statistics, produced by IMF
IMF	International Monetary Fund, Washington, DC
ITA	International Tin Agreement
ITC	International Tin Committee (to 1946); International Tin Council
ITRDC	International Tin Research and Development Council
ITRI	International Tin Research Institute

ITS	International Tin Statistics, produced by UNCTAD
ITSG	International Tin Study Group
KLTM	Kuala Lumpur Tin Market
KLSE	Kuala Lumpur Stock Exchange
LDC	less developed country
LME	London Metal Exchange
LTC	London Tin Corporation
MB	*Metal Bulletin*, daily, London
MBM	*Metal Bulletin Monthly*, London
MMC	Malaysia Mining Corporation
MNC	multinational corporation
MT	*Tin/Malaysian Tin/Malaysian Tin Bulletin*, monthly, Kuala Lumpur
NEI	Netherlands East Indies (colonial Indonesia)
ODI	Overseas Development Institute, London
OMO	Offshore Mining Organisation, Thailand
OPEC	Organisation of Petroleum Exporting Countries
RM	*Ringgit Malaysia* – new symbol for the Malaysian currency, announced 1992, which was previously abbreviated as *M$*
SOE	state-owned enterprise
SSB	State Statistical Bureau, People's Republic of China
TFS	tin-free steel (steel plated with chromium)
TI	*Tin International*, quarterly magazine, London/Kuala Lumpur
UMS	Unfederated Malay States
UN	United Nations
UNCTAD	United Nations Conference on Trade and Development
UNIDO	United Nations Industrial Development Organization
WBMS	World Bureau of Metal Statistics

CHAPTER I

Introduction to the World Tin Economy

Contrary to popular belief, industrial countries do not depend mainly on the Third World for their mineral supplies.[1] Among the top ten (non-fuel) mineral exporters, only two (Chile and Brazil) are less developed countries, and they are ranked only ninth and tenth, respectively (UNCTAD, 1992a). For those less developed countries that do possess minerals, though, mineral exporting has proved a mixed blessing. During the colonial period, mineral exports were associated with the dominance of multinational mining companies, and the benefits of the sales of minerals were often not channelled back into the economies which produced them. Political independence has not necessarily produced better results. Many state mineral companies, set up after foreign companies had been nationalised, are strikingly inefficient. Foreign investors have been discouraged by unfavourable conditions, although attitudes towards foreign investment are now changing. In addition, where a less developed country is heavily dependent on minerals, mineral exports may distort the economy and hamper other development. Even industrial countries have not been immune from the adverse effects of mineral exports: witness the British economy in the early 1980s, where recession and the wiping out of much of the manufacturing sector can be traced to the overvaluation of the pound, brought about in part by the growth of North Sea oil exports.

Tin is unusual among minerals in that the world *is* dependent on less developed countries for the bulk of its supply. It is also unusual in that tin exports in several of the largest producers appear to have had quite positive effects on those countries' development. The tin industry includes producers which have progressed far beyond mineral dependence, like Malaysia, and poor, highly mineral-dependent economies such as Bolivia. One theme of this book will be why tin has tended to promote development, whereas many other minerals have not, and why its effects have been stronger in some countries than others.

In 1985 the tin industry experienced the spectacular failure of the International Tin Agreement, and the collapse in prices which followed has continued into the 1990s. In that time the structure of the industry has changed, and the long-term future of many traditional producers has been threatened.

This chapter first describes further the interest of tin as a commodity study, and sets out in more detail the main themes of the book. It then discusses the pattern of tin production, consumption and trade, and looks at the tin market, the International Tin Agreements and the US strategic stockpile of tin. It describes the different tin-mining technologies and the mining industry's cost structure, before looking at the economic organisation of the industry. It also considers mineral development policy and ways of analysing the economic development effects of tin exports. Finally, introductions to the major tin exporters (Malaysia, Indonesia, Thailand, Bolivia, Brazil and China) are given.

TIN AS A COMMODITY STUDY

In the past, the tin industry has attracted attention because of the longevity and apparent success of its international commodity agreement, the International Tin Agreement. Since the bulk of tin production is found in a small number of less developed countries, while consumption, like that of other metals, is mainly by industrial countries, attention has also focused on the industry's cartelisation potential. This potential, however, has been constrained by the fact that since the 1950s the US has maintained a strategic stockpile of tin metal equivalent to one to two years' world production.[2]

The industry has been of interest to minerals economists because of its wide range of production costs between countries, which gave rise for many years to substantial profits for intra-marginal producers. Bargaining over these profits between host governments and foreign investors raised issues of mineral taxation policy, of government equity participation in mineral ventures, and of the role of state ownership and local companies in relation to foreign investment. Yet tin has been different from other major non-ferrous metal industries; the large international mining companies which tightly controlled those industries were largely absent from tin until the late 1960s, and tin-mining companies were specialised. Also, many tin-producing countries have significant, local small-scale tin-mining sectors, and the effects of these sectors on the countries' economic development have often been very favourable. Even foreign investment in tin in the past seems to have generated considerably more economic development for host countries than that in other minerals.[3]

From the mid-1970s the world mining industry for most metals experienced a prolonged recession (Bomsell *et al.*, 1990; Mikesell and Whitney, 1987; and Crowson, 1991). Tin was not immune from these problems, but its own severe slump was postponed until 1985. The collapse of the International Tin Agreement in October of that year was followed by a halving of the world price to below the production costs of many previously low-cost mines in South-East Asia, the then centre of the world tin-mining industry. Indeed, tin experienced the largest price fall of any major primary commodity during the 1980s, a period in which many primary commodities experienced historically low prices (Maizels, 1992). Simultaneously during the 1980s, Brazil moved from being a minor producer to the world's largest, with large reserves and very low costs, challenging the previous dominance of Malaysia, Thailand and Indonesia. Following rancorous negotiations on the setting up of the Sixth (and last) International Tin Agreement, Malaysia in 1981–2 tried unsuccessfully to manipulate the tin price on the London Metal Exchange, incurring substantial losses in the process and generating difficulties for the ITA which would contribute to the Agreement's downfall three years later. Brazil has remained outside the Association of Tin Producing Countries, set up by the major producers in 1983 while the ITA was running into difficulties. China too has remained outside the ATPC, though indicating an intention to join; it has entered world markets as a seller while ATPC members have been restricting output, and in 1990 produced a larger output even than Brazil's.

In presenting a study of the tin industry, this book will adopt similar themes to other volumes in this commodities series, and in chronological sequence. The themes include: the patterns of trade, demand and consumption; the economic structure of production and processing; the international regulation of tin; technical change; financing; the labour process; and the role of the state. Different chapters, however, will not give equal weight to the various themes. At different periods through history, and in different countries, different issues dominate. Thus a discussion of the late nineteenth century can hardly fail to give heavy emphasis to the rapid growth of what was then British Malaya, and the competition in that country between western mining companies and ethnic Chinese miners, a competition conducted to a significant extent through technical change. Similarly, the work on the period immediately after the second world war will stress the Bolivian and Indonesian nationalisations. The chapter on the 1960s and 70s will concentrate on producer countries' policies with regard to improving the host-country gains from foreign investment. That chapter will include a case study of the effects of tin exports on development, when those effects were at their greatest, and when there were substantial mineral rents to be bargained for.[4] When the book moves to the 1980s and beyond, interest in

TABLE 1.1   Production of tin-in-concentrates, by country, 1801–1990

(000 tons /tonnes)	1801–10	1861–70	1913	1929	1950	1965	1980	1990
Malaysia	3.3	8.5	51.4	72.3	57.5	63.7	61.4	28.5
Thailand	–	–	6.7	9.9	10.4	19.0	33.7	14.4
Indonesia	0.4	6.0	20.9	35.9	32.1	14.7	32.5	31.1
Bolivia	–	0.1	25.9	46.3	31.2	23.0	27.3	17.2
Brazil	–	–	–	–	0.2	1.8	6.9	39.1
China	2.0	0.5	8.3	6.8	7.5	25.0	16.0	44.0
UK	3.1	9.4	5.3	3.3	0.9	1.3	3.3	4.2
Nigeria	–	–	4.0	11.1	8.3	9.5	2.7	0.2
Zaire	–	–	–	1.0	13.5	6.3	3.2	1.6
Australia	–	0.1	7.8	2.2	1.8	3.8	11.6	7.4
South Africa	–	–	2.3	1.2	0.6	1.7	2.9	1.1
USSR	NA	NA	NA	NA	c. 8.9	23.0	16.0	15.0
World	9.0	25.0	134.0	196.0	169.3	201.1	235.9	220.6

Sources and Notes
1. 1801 to 1929 figures from ITSG (1949), and are shown in long tons.
2. 1801–10 and 1861–70 are annual averages, and figures for those years for 'Malaysia' are for Federated Malay States and Thailand.
3. 1950 and 1965 figures are from Baldwin (1983, pp. 12–13) and are in long tons.
4. 1980 and 1990 figures from UNCTAD (1992a), in tonnes.

tin and economic development in the old established producer countries will give way to a consideration of what are the tin-mining industry's chances of survival. Each of the chapters will conclude with a short overview of the period it has covered.

THE PATTERN OF TIN PRODUCTION, CONSUMPTION AND TRADE

Tin is one of the earliest metals known to the human race. It appears to have been in use before 3000BC in parts of Europe and Asia. Its property of hardening copper, to form the alloy bronze, made it of military significance since bronze weapons were superior to those made from copper alone. Similarly, tools made of bronze were preferred. Bronze, which contained about 10 per cent tin, could be moulded more easily than copper to make better castings. The widespread demand for tin during the bronze age, and the fact that tin deposits were not similarly widely scattered, made tin an important item of trade (Hedges, 1964, ch. 1).

The use of tin in tinplate (steel coated with tin), until very recently tin's major use in most consuming countries, dates back to the beginning of the nineteenth century when the value of tin's non-toxic, non-corrosive properties in the preserving of food in tinplate cans was first recognised (Hedges, 1964, ch. 10). As Table 1.1 shows, by the first decade of the twentieth century world output of tin exceeded 100,000 tons, about half present day

TABLE 1.2   Tin production and exports of the ten largest producers, 1990

(000 tonnes)	Tin-in-concentrates production	Primary tin metal production	Tin metal exports	Net exports [+] or imports [-] of tin-in-concentrates
China	44.0	28.0	10.1	15.7 [+]
Brazil	39.1	35.1	27.6	1.7 [+]
Indonesia	31.1	31.0	29.0	0.3 [+]
Malaysia	28.5	49.0	52.7	21.4 [-]
Bolivia	17.2	13.4	13.7	3.7 [+]
Former USSR	15.0	16.0	–	–
Thailand	14.4	15.5	11.9	0.1 [-]
Australia	7.4	0.3	0.2	8.1 [+]
Peru	5.2	–	–	4.6 [+]
UK	4.2	6.1	6.8	2.7 [-]
World	220.6	213.7	193.0	44.9 [gross exports]

Source: UNCTAD (1992a)

output. By the first world war virtually all the major producing countries of recent years were already significant producers, except for Zaire, and of course, Brazil. By 1929, the last year before prolonged recession, output (196,000 tons) was almost at present day levels; and the five largest producers, all less developed countries, produced some 90 per cent of world tin output, almost entirely for export. These, in order of importance, were: Malaysia, Bolivia, Indonesia, Nigeria and Thailand, all (with the exception of Nigeria) important players today.

In 1980, the basic pattern of supply and demand for tin remained that which had been in place for most of the century; world tin output came mainly from a small number of less developed countries, who exported the bulk of it to industrial countries. Among developed countries, only Australia and South Africa were net exporters, Australia's tin production having risen since the 1960s as part of a more general mining boom (McKern, 1976). The two largest communist countries were both large producers and large consumers. China produced a small export surplus, and the then USSR was a net importer. Malaysia, the world's largest producer, had been so since 1879 (when it overtook the UK and Australia), except for the second world war when it was occupied by the Japanese.

Although the 1980s have seen great changes in the industry, the overall dependence of the world economy on tin production from the Third World has not been altered. Developed market economies still produce well under 10 per cent of world output. Less developed countries produce some two-thirds, and this rises to nearly 85 per cent if China is added. Within the

TABLE 1.3   Shares of less developed countries in tin and other world non-fuel mineral exports, 1990

	World exports (US$ mil)	Percentage share of less developed countries in world exports	Percentage share of less developed countries + socialist countries of Asia in world exports
Bauxite	998.0	83.0	87.4
Alumina	5,586.6	27.3	28.0
Aluminium	**154,248.2**	**2.4**	**2.7**
Copper ore	3,799.1	48.7	48.7
Unrefined copper	1,449.5	79.6	79.6
Refined copper	**9,726.1**	**55.3**	**55.3**
Iron ore	**8,623.9**	**45.9**	**45.9**
Lead ore	627.2	28.3	28.3
Lead metal	**1,129.4**	**23.1**	**26.6**
Manganese ore	721.3	39.9	40.2
Ferromanganese	**750.4**	**11.5**	**16.4**
Nickel ore	200.2	91.0	91.0
Nickel intermediate products	**1,261.2**	**42.8**	**42.8**
Unwrought nickel	457.0	24.4	24.5
Phosphate rock	**1,422.4**	**62.7**	**63.5**
Sulphur	**1,494.7**	**23.3**	**23.3**
Tin ore	204.0	38.2	63.7
Tin metal	**1,225.8**	**84.6**	**89.9**
Tungsten ore	**93.2**	**19.5**	**80.8**

Sources and Notes
1. All figures calculated from UNCTAD (1992a).
2. Socialist countries of Asia are China, Mongolia and Vietnam.

Third World too production is quite concentrated. The top five producers (see Table 1.2), all LDCs (if China is counted as an LDC), account for 72 per cent of world output. However, the rise of Brazilian production from under 7000 tonnes in 1980 to over 50,000 tonnes in 1989 (UNCTAD, 1992a) is striking, as also is the growth of China's output, which in 1990 overtook Brazil's. This has reduced the relative importance of South-East Asia, where the output of Malaysia and Thailand has fallen significantly in the

1980s. Malaysia was overtaken by Brazil as the world's largest producer of tin ore in 1987. The importance of Bolivia, a high-cost producer much damaged by the 1985 price collapse, has also declined.

In spite of some growth in the domestic use of tin by LDCs, most of their production is still exported (see Table 1.2). Indeed, less developed countries had a higher market share in tin exports than in almost any other major mineral. In Table 1.3 only nickel ore shows a higher LDC share, but the share of LDCs (+ China and other Asian socialist countries) is much higher in tin overall i.e. ore + metal (86 per cent) – than in nickel overall – i.e. ore + intermediate products + unwrought nickel (44 per cent).

The production figures in Table 1.1 refer to *tin-in-concentrates*, which is the assumed tin metal content of ore concentrated at the mine prior to being sent for smelting. Tin has carried further than most minerals the tendency for LDCs to process their ore into metal. This is clear from Table 1.3 where the LDCs' share in other minerals tends to be lower the higher the stage of processing, most dramatically so in bauxite/aluminum. Thus most tin enters world trade as tin metal, and 1990 world exports of tin ore were an amount equivalent to only 17 per cent of tin metal exports. Only 20 per cent of world tin-in-concentrates production is exported, compared to 90 per cent of tin metal production.

Domestic smelting of tin ore dates back to the nineteenth century in the case of Malaysia. Other major producers tended to follow the colonial pattern of ore being shipped to the parent country for smelting, although until Thailand set up its own smelter in the 1960s, Malaysia smelted much of Thailand's output. The Netherlands and the UK both maintained substantial smelting capacity into the post second world war period. In the 1960s and 1970s there were great increases in smelting capacity in the Third World, as Thailand, Indonesia and Bolivia established smelters. Bolivia has had difficulty in smelting all of its own ores, basically because they are complex and hard rock, whereas those of South-East Asia are alluvial (see below). Bolivia has reduced the proportion of its mine output which is domestically smelted since 1985 while its tin industry has been being re-organised. In alluvial tin-mining ore is generally concentrated at the mine to 70–5 per cent purity, so smelting involves a weight loss of some 25 per cent, while costing in the early 1980s (i.e. before the price collapse) only a few per cent of the value of output. Malaysia is now the world's largest importer of tin concentrates (tin ore), maintaining the throughput of its two smelters in the face of drastically reduced domestic mine output. Brazil smelts most of its domestic tin mine production; indeed Brazil's smelting industry antedates its mining of tin in significant quantities. Some Brazilian ore is exported illegally to Bolivia though, mainly by small-scale miners and in exchange for drugs. China exports about a third of its production of tin

TABLE 1.4   Tin metal consumption, 1990

(000 tonnes)	Primary tin metal	Secondary tin metal
USA	36.6	7.7
Japan	34.8	5.1
Former USSR	26.0	3.0
Germany	18.7	0.3
China	17.0	–
Brazil	9.0	–
France	8.3	–
South Korea	7.4	0.6
Italy	6.9	–
UK	6.5	3.9
World	**228.9**	**22.8**
Developed market economies	134.3	19.0
Less developed countries	42.1	0.6
Eastern Europe	35.5	3.1
Socialist countries of Asia	17.0	–

Source: UNCTAD (1992a)

in the form of tin concentrates, about a third of which goes to Malaysia, which also smelts the bulk of Australian tin output.

As Table 1.4 shows, developed countries still account for nearly 60 per cent of world consumption of primary tin, and almost all of that of secondary (i.e. recycled) tin. Recycled tin constitutes only a small proportion of world consumption (9 per cent in 1990 (UNCTAD, 1992a)) since the tin content of tinplate (the main source of recycled tin) is low. The proportion of tin consumption generated by developed countries is less than a decade ago (65 per cent in 1980), as tin producers and other LDCs have developed tin-using industries of their own, especially tin-plating. Brazil is now one of the top ten world consumers of tin.

Tin-bearing ores are associated with granitic rocks, and the only ore which is mined to yield an economic return is *cassiterite*. This is found either as *primary deposits* in the form of *lodes*, or as *placer deposits*. Lodes are normally thin veins, running through hard rock. Placers are accumulated by the breaking down of the granitic rock over a long period of time. Placers are most usually found as *alluvial* deposits, where the primary tin deposit has been broken down by the action of water and occurs fairly near the surface in river valleys and sometimes offshore. They may also occur in *eluvial* deposits, where the primary material has been decomposed, but has not been transported any distance by water.[5]

Lode deposits make up 44 per cent of world reserves, as shown in Table 1.5, and placers the rest (Sutphin *et al.*, 1990, p. 1). Probably the world's richest

TABLE 1.5   World tin reserves

(000 tonnes)	Total Reserves	Placer Deposits	Lode Deposits
China	1,562.5	NR	1,562.5
Malaysia	1,207.6	1,200.0	7.6
Brazil	1,195.5	1,195.5	NR
Thailand	938.4	864.7	73.7
Indonesia	821.3	798.1	23.2
Zaire	510.0	NR	510.0
Bolivia	453.7	12.8	440.9
Former USSR	300.0	NR	300.0
Australia	207.8	NR	207.8
Namibia	120.8	NR	120.8
Nigeria	110.8	110.8	NR
Canada	92.4	NR	92.4
Portugal	72.8	NR	72.8
Japan	41.7	NR	41.7
Zimbabwe	31.7	NR	31.7
South Africa	24.2	NR	24.2
Peru	22.5	NR	22.5
USA	2.4	NR	2.4
Burma	0.4	NR	0.4
Argentina	NR	NR	NR
UK	NR	NR	NR
Total	7,716.6	4,181.9	3,534.7

Source and Notes
1. Figures from Sutphin *et al.* (1990, p. 14).
2. The figures are for reliable estimates from identified deposits having economically expoitable resources.
3. No recorded entries are given for Argentina and the UK.
4. Figures for Brazil are almost certainly an underestimate (see text).
5. NR = none reported.

deposit of tin is the Asian tin belt which runs from Indonesia, through Malaysia, Thailand, Laos and Burma, into the southern Chinese province of Yunnan (Bureau of Mines, 1987, p. 268). China's deposits are mainly lodes, known for over 2000 years, while the tin in the rest of the belt occurs in placers.

By the 1970s tin was regarded as one of the few minerals in genuinely short supply (Bosson and Varon, 1977, p. 59). The reserve figures shown in Table 1.5, when compared to world production figures from Table 1.2, suggest that existing reserves have a life of only 34 years. The reserves are figures from the US Geological Survey and refer to reliable estimates of known deposits which are economically exploitable (Sutphin *et al.*, 1990, pp. 2–3) This estimate is larger than some others, for example that cited in

FIGURE 1.1   Real price of tin, 1860–1991 (1980=100)

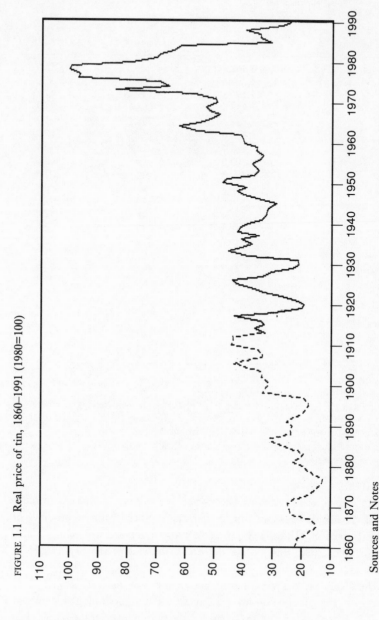

Sources and Notes

1. 1880–1914 is Straits tin price, from Wong (1966, p. 243) deflated by Sauerbeck–Statist index of British wholesale prices from Mitchel and Dean (1982, p. 474). 1914–91 is New York tin price, from Yip (1969, pp. 395–6) and *TI*, deflated by US wholesale price indices from Wattenberg (1978, p. 200) and IMF (1992).

2. Straits tin price index spliced into New York tin price index assuming same real price for 1914.

Crowson (1992b, p. xiii), which gives a reserve life of tin of only twenty-eight years, which is lower than any major mineral except for silver and lead. Brazilian reserves are almost certainly understated in these figures, and Sutphin *et al.* (1990, p. 14) cite a more recent estimate of 4 million tonnes, compared to the approximately 1.2 million tonnes shown in Table 1.5. The Patinga deposit in the Amazon, first worked in the 1980s, alone is thought to have reserves of 575,000 tonnes of tin (Sutphin *et al.*, 1990, p. 16), and the Bom Futuro deposit developed in the Amazon region in the late 1980s may be even larger.

Reserve figures, however, should be treated with great caution, and many past estimates have been highly inaccurate. The US government's Paley Commission in 1952 estimated Malaysia's tin reserves at less than the cumulated annual production from then to the late 1970s (Page, 1977, p. 5.5). The Fermor report on the mining industry of Malaya in 1939 gave tin reserve figures for the country of only 610,000 tons (Fermor, 1939, p. 119), half that of present-day reserves. Even long-established producers such as Malaysia have by no means prospected their land area fully, and the same is especially true of Brazil. 'Reserves' depend in any case on price and technology, and there have been long-term reductions in the real costs of extraction of minerals (Barnett and Morse, 1963; Barnett, 1979). Reserves give only a snapshot view of the current availability of a mineral, and the ultimately recoverable amount is likely to be much greater (Sousson, 1988, p. 9). Indeed, if one considered the elemental resource base for tin, in the sense of the amount available in the top mile of the earth's crust, there would be enough tin (and many other minerals) for millions of years (Rees, 1990, pp. 17–19). What does seem likely from published reserves figures is that dependence on Third World sources will continue.

THE MARKET FOR TIN, THE INTERNATIONAL TIN AGREEMENTS AND THE UNITED STATES' STRATEGIC STOCKPILE

The apparent scarcity of tin which had emerged by the 1970s, mentioned in the previous section, led to tin experiencing a rising price in the 1960s and 1970s in real terms (i.e. allowing for inflation), as Figure 1.1 shows.[6] Over the long term, tin's price has risen considerably in relation to that other major metals.[7] By the early 1980s tin was widely regarded as a semi-rare metal (Baldwin, 1983, p. 8). Compared to other non-ferrous metals it was produced in small quantities and its price stood between the prices for those metals and that for the precious metals (see Table 1.6). Since the collapse, the price of tin compared to other non-ferrous metals has weakened. Although the mid-position of tin has been maintained, except relative to nickel, the ratio of the price of other non-ferrous metals to that of tin has risen several fold. Nickel has become more expensive than tin, having been

TABLE 1.6   Outputs and prices of tin and major metals, 1980 and 1990

	1980 output (000 tonnes)	1980 price ($ per tonne)	1990 output (000 tonnes)	1990 price ($ per tonne)
Aluminium	16,051	1,728	17,987	1,639
Copper	9,307	2,173	10,684	2,661
Zinc	5,806	762	7,097	1,519
Lead	3,412	907	3,057	811
Nickel	739	6,520	874	8,864
Tin	232	16,785	214	6,235
Silver	10.6	661,662	11.3	154,966
Gold	1.2	19.7 million	1.7	12.3 million

Sources and Notes
1. All figures are from UNCTAD (1992a); except for zinc, gold, and silver production figures for 1980 which are from Baldwin (1983, p. 9), and for 1990 from Crowson (1992b).
2. Prices are free market, London Metal Exchange prices.
3. Gold and silver prices are converted from $ per troy ounce.

not much more than a third of the price of tin in 1980.

More generally, although primary commodities in general experienced exceptionally historically low prices over the mid-1980s (Maizels, 1987 and 1992), tin's price fall was greater than for the others. Over the period 1980–91, UNCTAD statistics record that tin experienced the largest annual average price fall of all 43 principal non-fuel primary commodities listed in the *Commodity Yearbook*. From 1980 to 1991 tin's real price declined by an average of 13.1 per cent a year, compared to a 1.3 per cent average for the ten minerals in the *Yearbook* (UNCTAD, 1992a).[8]

There has been a long-term problem of slow growth in tin consumption, even though the trend price of tin did not rise significantly for the whole of this century until the 1960s (Humphreys, 1982; and Figure 1.1). The problem was intensified by the rising real price in the 1960s and 1970s. From 1900 to 1972 tin metal consumption rose only 2.2 times, compared with 4 times for lead, 6 times for zinc and 12 times for copper (Fox, 1974, pp. 17–18). From 1965 to 1985 world consumption of tin per unit of gross national product (GNP) halved, compared to a fall in copper consumption of about 25 per cent and a rise in the consumption of aluminium of about 20 per cent (ODI, 1988). In the 1970s world tin consumption declined by an annual average of 0.4 per cent. The decline in demand seems to have been halted in the 1980s, however, with the falling real price of tin. There was a positive annual growth in consumption of 0.5 per cent over the decade as a whole (Crowson, 1992b, p. 262), influenced by macroeconomic upturn in the late 1980s and a strong increase in consumption in the Third World. By the early 1990s, however, the collapse of the economy of the former USSR led to the loss of an important source of demand. Although real tin prices were falling in the 1980s and 1990s, it is not necessarily the case that increases in consumption can be associated in the short run with lower tin prices. Econometric studies generally have failed to identify significant short-run demand elasticities, probably because in tin's main uses it takes time to adjust inputs in reponse to a change in input prices (World Bank, 1990, p. 129).

Tinplate consumption showed sharply negative growth in the 1980s in the major industrial countries, and by 1986 solder had overtaken it as the most important use, generating 30.1 per cent of total demand compared to 28.1 per cent for tinplate (UNCTAD, 1990, p. 14). The other main uses of tin are to make tin alloys (for example for bearings) and tin chemicals, particulary for PVC stabilizers. Of the Western world's tin consumption in 1990, 31.4 per cent was used to make solder, 28.9 per cent for tinplate, 14.4 per cent for tin alloys, and 12.4 per cent for tin chemicals (Frame, 1992, p. 10).

Tinplate demand has been much affected by technical substitution, by competition from other packaging materials such as aluminium, plastics and glass, and by alternative means of food preservation such as freezing. Electrolytic tinplating has greatly reduced tin consumption per unit of tinplate compared with the older hot-dip process used before the second world war. By the late 1970s a food tin can was using only 20 per cent of the tin (and 65 per cent of the steel) used in 1945 (*TI*, December 1978). Nor does the structure of the world tin industry (discussed later) predispose tin users towards any vested interest in the continued use of tin, since tinplate producers are overwhelmingly makers of steel. Japanese manufacturers in the late 1960s developed chromium-plated steel (*tin-free steel*, TFS) as a competitor to tinplate. One large American can-making company is on record as trying to free itself from dependence on foreign supplies of tin (Robertson, 1982, pp. 92–4). Solder's increased importance reflects the world growth of the electronics industry.

In many metals large international mining companies have been able to exercise considerable control over markets, keeping prices high and stable through a system of published producer prices; free markets often have cleared only residual supplies. Over the 1960s and 1970s structural changes occurred which weakened this control, as new entrants such as oil companies, state-owned enterprises and consumer-led consortia increased the heterogeneity of producers (Bomsell *et al.*, 1990; Auty, 1987). Since the late 1970s the system of producer prices has weakened and free market prices have increased in importance (Slade, 1989). Compared to metals such as copper or aluminium, companies in tin generally have been far less able to exercise such control, except indirectly through influencing the setting up of commodity agreements (see Chapter 3). Tin has had a history of being traded in free markets on an arms-length basis. The main markets are the London Metal Exchange, the Kuala Lumpur Tin Market in Malaysia, and the tin market in New York. Before 1984, Malaysian tin was traded at Penang. However, the use of long-term contracts has increased considerably since the 1970s with the growing importance of state owned agencies in the major producers. For example, P. T. Timah, the Indonesian state tin company, markets half its output on long-term contracts (to Japan) (Sudarsono, 1988).

Tin, though, has had a long history of control by international agreement. Tin exports were first restricted by the so-called Bandoeng Pool in the early 1920s, an agreement among the colonial governments of Malaya and the Netherlands East Indies (Indonesia) following a sharp fall in the world price. A wider scheme of control was agreed in 1931 between Malaya, the Netherlands East Indies, Bolivia and Nigeria, using export quotas. A second agreement was set up in 1934 and a third continued until the start of the second world war. Buffer stocks also were used by the agreements in order to control price fluctuations. In the postwar period, starting in 1956, six International Tin Agreements have operated, with the aim of controlling fluctuations to within a specified floor and ceiling range, by means of purchases and sales of tin from the ITA's buffer stock backed by export controls when necessary. The sixth agreement became inactive after the tin collapse of 1985, and the International Tin Council was finally wound up in 1990.

The United States strategic stockpile of tin has also been a significant influence on the market. Stockpiling started early in the second world war, reaching 100,000 tons by 1942, but the major build-up occurred in the 1950s, reaching probably about 350,000 tons in 1957. This size, equivalent to seven years' US consumption, has been reduced by controlled releases since 1962. By 1974 over 140,000 tonnes had been released, including 40,000 during the 1973–4 commodities boom. The US agreed with the International Tin Council in 1966 to undertake stockpile disposals in such a way as to avoid market disruption (Fox, 1974, chs 11 and 15), but even today prospective releases from the stockpile are a worry for producers.

TIN MINING TECHNOLOGY AND COSTS
Introduction

This section introduces the main methods used to mine tin. A knowledge of technology is not only useful background information, but is necessary for an understanding of the economic organisation of the industry. It is also important for an understanding of the industry's development effects.

Mining techniques for tin broadly can be classified according to whether the deposits to be mined are primary deposits or placer deposits. In the tin industry, placer deposits are worked with surface methods, particularly gravel pumping and dredging. Both methods are hydraulic mining, where the use of water is central. Primary ('hard rock') deposits are normally worked by underground mining, although dry, open-cast mining is sometimes used. Prior to the 1985 tin collapse, the International Tin Council published detailed statistics on a (non-communist) world basis on tin production by mining technique,[9] and a set of these are given in Table 1.7 for 1979. These show gravel pumping to have been the most important single mining method, followed by underground mining, and dredging. The importance of underground mining will have been understated by the exclusion of China and the then Soviet Union from the statistics. The rise of Brazil in the 1980s will have increased the importance of alluvial mining methods, although Chinese output, which is mainly from underground mines, has greatly increased too. *Tin International*, the tin industry's trade magazine, estimated (in July 1988) that 53 per cent of the (then non-communist) world's producing and undeveloped deposits were workable by gravel pumping, 28 per cent by dredging, 15 per cent by underground mining and 4 per cent by open cast mining.

The deposits of South-East Asia, as reference back to Table 1.5 (on reserves) shows, are placers, as are those of Brazil, Nigeria and Zaire. Gravel pump mining and dredging have been the dominant methods in South-East Asia since the early years of this century. They are basically similar as between Malaysia, Thailand and Indonesia, though the latter two countries had offshore as well as onshore dredging, and Malaysian mines in general are more technically advanced in both methods. Historically, especially in Malaysia and Thailand, dredging has been associated with Western foreign investment, and gravel pumping has been run by local, ethnically Chinese miners. Brazilian alluvial mining methods make a somewhat greater use of dry mining techniques, and some different methods of dredging. Dry, open-cast mining is also much used in Nigeria and Zaire. These differences compared to South-East Asia largely reflect differences in the type of placer deposits, those in Brazil being more ancient and variable (Hosking, 1988, p. 45). In some respects, though, Brazilian methods embody technical progress which has influenced South-East Asia in the 1980s and 1990s as costs have needed to be cut to cope with price decline. Among past and present major producers, underground mining characterises Bolivia, China, the former USSR, the UK, and also Australia to a large extent.

As Table 1.7 shows, the different methods of mining differ greatly in terms of output per mine, with gravel pump mines producing on average

TABLE 1.7   Tin mining methods, 1979

	Percentage of (non-communist) world output	Number of units operating	Tones of tin-in-concentrates per unit per year
Gravel pumps	28.5	1304	44
Underground mines	22.0	33	1333
Dredges – onshore	13.2	90	295
Dredges – offshore	6.5	24	547
Suction boats	7.5	*c.* 2000	7
Dulang washing	1.9	*c.* 27500 (workers)	0.1
Opencast mines	1.7	23	152
Minor methods not reported to ITC	18.7	–	–

Source: Thoburn (1981a, p. 41)

much smaller outputs than dredges or underground mines. The figures reinforce the point made at the beginning of this chapter that small-scale mining is important in tin.

This section first introduces each of the main mining methods. It then considers costs, the economics of the choice of technique, and technical progress.

Mining methods

Gravel pumping

Gravel pumping is open pit mining, where the tin ore in mined from the sides of the pit by being broken down by high-speed jets of water played from rigid hosepipes ('monitors'), each operated by one or two workers. The 'slurry' (liquid ore and mud) is sucked up from the bottom of the pit by pumps ('gravel pumps') and elevated onto a long downward sloping sluice box (or 'palong', as it is known in South-East Asia) for primary concentration, after oversize material first has been screened out. Since the specific gravity of cassiterite is much greater than the material in which the tin is found, the separation of the tin ore from the slurry can be achieved by allowing the tin to sink to the bottom of the palong, while the waste material runs off. Jigs can also be used for this primary treatment, and in Brazil this is the normal method. These are pulsating boxes in which the waste material is floated off the ore, and they often are used for further concentration at the mine head if a sluice is in operation. In South-East Asian operations, the ore is concentrated at the mine head to 15 to 30 per cent purity (the percentage depending on the risk of theft), and is further concentrated in a special plant (the 'tin shed') to about 70 per cent purity, where various by-products are also removed prior to shipment to the smelter. In Brazilian gravel pump mines the percentages would be about 20 per cent and 60 per cent, respectively. Using multi-stage pumping, a gravel pump mine can operate to a depth of 150 feet, which is as deep as a dredge, or sometimes even 200 feet, but they are usually much shallower.

Tin dredging

Tin dredging is carried out normally by either a bucket ladder dredge or an underwater bucket wheel dredge. A bucket ladder dredge, overwhelmingly the main dredging method in South-East Asia, is a large ship-like floating factory which works on a inland pond (or 'paddock'), mainly of its own making, and digs the tin-bearing ground from the bottom. The dredge is moved by means of lines fixed to the shore and attached to winches on the dredge. It is usually powered by electricity, but in remote areas diesel may be used, and steam power has been used in the past. Dredges have their own primary treatment plant on board, using jigs. The dredge enlarges its paddock as it mines, and dumps behind it the waste material ('tailings') from the treatment plant. The digging is done with a continuous line of steel buckets attached to a chain mounted on a metal ladder, which can be raised or lowered to alter the digging depth. Bucket ladder dredges can work to a depth of about 150 feet, and deeper if the level of paddock water is lowered, as in the famous Kuala Langat deposit in the state of Selangor in Malaysia. Dredges in Thailand and Indonesia also work offshore. Suction cutter dredges have also been used, though rarely. These suck ore from the bottom through a pipe attached to a revolving cutter. The best known is that used off the coast of southern Thailand by Billiton in the late 1970s and early 1980s.

Bucket wheel dredges, used in Brazil, float on a pontoon and dig with a set of open-bottomed buckets on a revolving wheel. The tin-bearing ground is sucked up through a pipe beside the bucket wheel, and is pumped to a separate, floating treatment plant which can follow the dredge. Bucket wheel dredges can dig only to about fifty feet, but many Brazilian deposits are shallower than those mined by Malaysian dredges, and the bucket wheel dredge's capital costs are much lower per unit of throughput.[10] Bucket wheel dredges can also be used to strip the barren *overburden* of earth from a deposit before mining with other methods. At present several bucket ladder dredges are also in use in Brazil, but they are still all of small capacity. Brazil also, however, has some larger very deep deposits which might be suitable for larger, bucket ladder dredges in the future (*MT*, October 1985, p. 8)

PLATE 1 Malaysian gravel pump tin mine.

PLATE 2 Malaysian gravel pump tin mine (close up).

PLATE 3 The palong on a Malaysian gravel pump mine.

PLATE 4 Tin concentrates delivered to the tin shed for further processing, Malaysian gravel pump mine.

PLATE 5 Tin ingots, Mentok smelter, Indonesia.

PLATE 6 Tin dredge, Malaysia, depositing tailings behind it as it digs.

PLATE 7 Offshore tin dredge, Bangka Island, Indonesia.

PLATE 8 Bucket ladder raised for maintenance, offshore tin dredge, Indonesia.

PLATE 9 Dulang washers mining for tin in Malaysia.

PLATE 10 Suction boat mining tin illegally off coast of southern Thailand in 1970s.

PLATE 11 Mining for tin with excavator, Rondonia, Brazil.

PLATE 12 Jig plant for primary treatment of tin ore, Rondonia, Brazil.

Washing Plants

Another Brazilian mining method is to use a floating treatment plant, like that accompanying a bucket wheel dredge but fed by an excavator from the bank. This method is also used with the excavators mounted on pontoons of their own, a method claimed to have been pioneered by Paranapanema, the largest Brazilian tin-mining company (Paranapanema, 1989)

Dulang Washing

This is the working of surface deposits by hand panning, like gold mining in rivers by prospectors in early America. The name comes from the pan, or *dulang* in Malay. Dulang washers are found in Malaysia and Thailand particularly, where in 1990 they accounted for approximately 9 per cent and 6 per cent, respectively, of total tin concentrates production.[11] In Malaysia at least, the ethnic Chinese women working as dulang washers have had an interesting form of social organisation; they live more or less independently of men, and continue their trade over the generations by adopting female children (Siew, 1961). Dulang washing has also been used by *garimpeiros* (small-scale miners) in Brazil.

Suction Boat Mining

Suction boat mining was a distinctive feature of the Thai tin industry, and started in 1975–6 off the country's south-west coast. Converted fishing boats used a suction pump operated on the sea bottom by a diver illegally to work the rich parts of leases allocated to large mining companies. These boats were still recorded as producing a quarter of Thai output in 1987 (DMR, 1990), but gradually have been replaced by mini offshore dredges, and are no longer recorded as producing anything in the official statistics. The boats were hard hit by the 1985 price collapse, and also by tougher government control and by the working out of the richer offshore deposits.

Open-Cast Mining

In the tin industry this refers to dry-mining, using excavators, draglines and similar earth-moving equipment to mine richer, more concentrated deposits. It can be used either to mine placer or shallow lode deposits. Probably the largest open-cast lode tin mine in the world is the Uis mine in Namibia (Bureau of Mines, 1987, p. 276),

In South-East Asia (dry) open-cast mining became rare after the first world war, though this is how the industry first developed, as Chapter 2 will show. This is in part because of the region's high rainfall and ready supply of water for hydraulic methods. It is a more common method in Nigeria and Zaire. In Brazil too, mines are more often a mixture of gravel pumping mining and dry-mining. This is not simply in the sense that

Brazilian mines make great use of earth-moving equipment to feed the monitors, as in modern South-East Asian mines, but that a particular mine may have several faces each operated by a different method. Excavators may directly feed dumper trucks, which take the ore to a nearby concentration plant. This occurs where the ground is too hard to be broken down easily by the monitors. In Brazil mining companies use more of a mixture of mining methods because deposits are more variegated than in South-East Asia. Fewer deposits in Brazil are purely alluvial; eluvial deposits and primary deposits broken down *in situ* are also found.

Underground Mining

Whereas most alluvial methods described here are more or less peculiar to the tin industry (except for some similarities to alluvial gold mining), underground mining to work lode deposits is basically similar to the familiar methods used to mine coal in Western countries. Veins of tin ore may be very narrow, however, and run for long distances through hard rock. The Gejiu mine in Yunnan in China is probably the largest in the world (*MT*, April 1986), and in the Western world the largest in Renison Bell in Australia. In the Renison mine, shafts run horizontally into a hillside, without vertical shafts (Robertson, 1982, p. 23). In Britain, the world's largest producer at times during the nineteenth century, tin was mined by underground methods. In Cornwall, the centre of British tin mining, only one mine now (1993) remains in production. In Bolivia, the underground mines are in the high Andes mountains, and the difficulty of mining conditions has made Bolivia one of the world's highest cost producers.

Costs, Choice of Technique, and Technical Progress in Tin Mining
Costs

Direct mining costs in alluvial tin mining depend principally on the throughput of ore-bearing ground treated, rather than the output of tin. The cost per unit of tin ore produced depends on the cost per unit of throughput in relation to the richness of the ground treated. The throughput is also the normal measure of the capacity of a mine, and is expressed in terms of cubic metres per time period.[12] For example, a large capacity tin dredge built in South-East Asia in the 1970s or early 1980s would have had a capacity of over 600,000 cubic metres a month (there has been virtually no large dredge building since then). A typical best-practice Malaysian gravel pump mine of the same period would have had a capacity of some 100,000 cubic metres a month; some would have been larger than this. There would also have been numerous much smaller producers, as in the other South-East Asian countries, but many of the smaller mines have gone out of business since 1985.

The richness (or *grade*) of the tin-bearing ground is expressed normally in terms of kilos of tin metal (or sometimes tin concentrates, containing 75 per cent tin) per cubic metre of throughput. During the 1970s, the International Tin Council produced a wide range of published statistics on grades, collated from information supplied by national mining authorities. In 1978, for example, Malaysian gravel pumping was operating with an average grade of 0.14 kg metal per cubic metre, and Malaysian dredges worked with an average grade of 0.11 kg metal per cubic metre (Engel and Allen, 1979, p. 84).[13] Grades in tin are much lower than for other non-ferrous metals, except precious metals. Taking a cubic metre of earth to weigh about a tonne, the Malaysian grades just cited would represent tin metal contents per unit of throughput of 0.014 per cent and 0.011 per cent. These compare to grades for copper of 0.4 to 6 per cent, zinc 4 to 16 per cent, lead 3 to 12 per cent and bauxite 15 to 25 per cent (Fox, 1974, pp. 5–6), which reinforce the perception mentioned earlier of tin as a semi-rare metal. Now, information on tin grades is fragmentary, but it is likely that the grades in Malaysia and other South-East Asian producers will be higher since only mines operating richer deposits will have been able to survive the low tin prices.[14]

Costs between alluvial tin-producing countries may differ greatly, therefore, according to their ore grades. Malaysian operations have been able to compensate for low grades compared to other South-East Asian producers (see note 13) by greater technical efficiency (Thoburn, 1981a, p. 124). The great difference between grades in South-East Asia and those in Brazil cannot be compensated for in this way, however. Brazilian gravel pump operations in 1979 (i.e. before the Patinga and Bom Futuro discoveries) were working with much higher grades, some 0.5 kg metal or more, although the cost advantage was offset to a some extent by the heavy infrastructural and personnel costs of operating in the remote Amazon region, and transporting the concentrates to the south for smelting (Engel, 1980). Grades in Brazil by the late 1980s often were in excess of 1 kg metal per cubic yard (interviews, Brazil, 1992). Grades on Paranapanema company's Patinga property, Brazil's largest mine, were reported in 1985 by a Malaysian mining mission to Brazil to be over 2 kg metal per cubic metre (*MT*, 15 October 1985). Garimpeiros (small scale miners) working on the huge new deposit at Bom Futuro in Rondonia in the late 1980s would have been working grades well over 3 kg. Garimpeiros use a variety of methods, sometimes including gravel pumps.

Until the rise of the Brazilian tin industry in the 1980s and the collapse of the tin price in 1985, the most significant cost difference among the major producers in the international tin economy was that between Bolivia and the South-East Asian operations. Throughput costs for underground mining are understandably considerably greater than for surface, alluvial mining. The higher grades associated with primary rather than alluvial deposits may compensate for these higher costs. In Bolivia the tin mines are located in the high Andes mountains, a difficult and inaccessible area. These natural disadvantages have been compounded by a long-term lack of exploration and investment in Bolivia, although Ayub and Hashimoto (1985, p. 39) argue that the high Bolivian costs in the late 1970s and early 1980s were exaggerated somewhat by an over-valued exchange rate.

These differences, particularly with the high real prices in the 1970s, gave rise to very large profits in tin mining in South-East Asia and in some operations in smaller producing countries. In 1974 and 1978 tin prices were high enough to give significant rents even to marginal producers (Ayub and Hashimoto, 1985, p. 33).

Table 1.8 shows cost data collected in a large-scale survey carried out by the US Bureau of Mines in 1982, expressed in US dollars in terms of 1985 prices. This gives an indication of cost in relation to the collapsed price of tin in late 1985. Following the tin collapse, with an average free market spot price for tin of US$2.85/lb in 1986, it can be seen from Table 1.8 that very little of the industry could cover even its direct operating costs. Of course, within each sector in each country there would be considerable spread of costs around the average, according to the richness of each mining property, but the basis for the drastic contraction of the industry in South-East Asia is clear. Although the late 1980s saw substantial rises in the tin price, the revival was shortlived and by 1991 in real terms the tin price was below that of 1986 (Crowson, 1992b, p. 254). The late 1980s also saw intense efforts to cut costs.

Choice of Technique

In addition to the large cost differentials between producing countries, and the profits these gave rise to in intramarginal producers before 1985, the other striking feature of the world tin industry is the range of production techniques operating in many of the producing countries. To some extent this is the result of the widespread existence of small deposits in these countries which allows small-scale methods to survive. This is true of minor techniques such as dulang washing and of the open-cast mining of very concentrated alluvial deposits. More important, however, is that among the main techniques, gravel pumping is often an alternative to dredging inland. For offshore mining, with shallower deposits which are close inshore, suction boats can substitute for offshore dredges. Thus, small-scale technology has existed as a vehicle for the entry of local mining firms. In Brazil too, the large, rich, shallow deposits can be worked by small-scale miners, and there has been intense conflict between them and the established mining companies.

TABLE 1.8　　Tin mining costs, 1982

(US$, at 1985 prices)	Net operating cost per pound tin metal	Recovery of capital per pound tin metal	Taxes and royalties pound tin metal	Total cost per pound tin metal	Mining and benefication cost per tonne of ore bearing material mined and treated
Gravel pumps					
Brazil	2.60	0.50	0.70	3.80	2.30
Indonesia	4.20	0.30	0.10	4.70	1.90
Malaysia	5.00	0.30	0.10	5.40	0.80
Thailand	4.50	0.10	1.80	6.40	1.50
Others	2.80	0.80	0.30	3.90	3.40
Average	4.70	0.30	0.30	5.30	1.00
Dredges					
Indonesia	3.80	0.80	0.20	4.80	1.10
Malaysia	3.20	0.00	0.50	3.70	0.40
Thailand	2.70	0.30	1.40	4.40	0.80
Others	4.30	0.50	0.30	4.80	0.90
Average	3.50	0.50	0.50	4.50	0.80
Underground					
Bolivia	4.40	0.40	2.20	7.00	42.30
South Africa	2.90	0.30	0.00	3.20	19.60
UK	4.30	0.00	0.20	4.50	45.20
SE Asia	1.80	1.20	0.20	3.20	46.90
Others	3.00	0.90	0.20	4.10	29.80
Average	3.40	0.80	0.60	4.80	34.20
Open Cast					
Australia	8.70	1.00	0.10	9.80	16.50
Thailand	2.90	0.50	1.30	4.70	10.70
Others	0.50	0.00	0.50	3.00	4.90
Average	4.10	0.70	0.10	4.90	9.90

Sources and Notes
1. Data from Bureau of Mines (1987, pp. 276–7)
2. 'Others' include: Australia, Bolivia, Burma and Zaire for gravel pumping; Australia, Bolivia, Brazil and Nigeria for dredging; Argentina, Australia, Japan, Peru and Zimbabwe for underground mining; and Brazil, Malaysia and Namibia for open-cast (but note Malaysia has little open-cast mining).
3. The average tin price in 1986 (i.e. after the tin collapse) was 42.85/lb. (Crowson, 1992b, p. 254).

The range of techniques is especially interesting in the cases of Malaysia and Thailand, where the main techniques (gravel pumping and dredging) have historically been associated with local as opposed to foreign ownership and control, and with somewhat different development effects. In the case of these two techniques, their coexistence to some extent can be explained in technical terms. There are certain areas which dredges cannot mine, such as hillsides, pinnacled limestone deposits, and in general any deposit of less than about 400 acres (necessary for minimum throughput). Conversely, gravel pump mines cannot operate in swampy areas or offshore, though they can operate (with double pumping) at the same depths (150 feet) as the deepest dredges. However, much ground worked by dredges could technically be worked by gravel pumps. The fact that gravel pumps are not working such areas may be due to the historical accident of the land being alienated to dredging companies as a result of preferential government treatment in the past, or it may be that dredging was the more profitable technique at the time (Thoburn, 1977b, p. 91). As I have shown elsewhere, the relative profitability of the two techniques may switch within the range of historically experienced prices. Higher prices (and also higher discount rates in an investment appraisal) favour gravel pumping, the less capital-intensive technique (Thoburn, 1977a, 1978a).

Technical progress

In the world mining industry generally, considerable technological change has occurred as larger scales of operation have been developed to deal with high-throughput/low-grade operations. Tin has experienced some of these changes and the grades operated have certainly declined in the main pro-ducers. In the same way that the size of trucks and mechanical shovels used in open cast hard-rock mining has increased several-fold over the years (Bosson and Varon, 1977, p. 34) the throughput of mines has increased in tin. However, in more recent years the rate of increase in the size of such earth-moving equipment has slowed (Ericsson, 1991).[15]

Before the second world war dredges typically used 7–14 cubic feet buckets and a monthly throughput of 200,000 cubic yards was quite large. By the 1970s 600,000 cubic yards had become common for new dredges and bucket size had increased to over 20 cubic feet (Abdullah and Chan, 1991). In the late 1970s further increases in size took place and one dredge commissioned at the end of 1979 had 30 cubic feet capacity buckets. Computerisation too has affected tin as well as other metals and by the late 1970s, after which little new large dredge building has taken place, the newer offshore dredges were using computer control. In underground mining too some changes in techniques have occurred, such as block caving to allow the mining of lower grades of ore. Equally important, the small-scale sectors of the tin industry have shown adaptability to lower grades. Gravel pump mines have increased their throughputs, particularly in Malaysia where low grades are more usual than in Indonesia or Thailand. Earth-moving equipment is used to drystrip and to feed the monitors, and more mechanised forms of ore treatment are used. Increased use of earth-

moving equipment, and some moves towards dry-mining have been intensified since the 1985 price collapse, which has left a much reduced number of
mines, generally ones which are both technically efficient and which have
richer deposits. This is discussed further in Chapter 6. Brazilian mines also
make much use of earth-moving equipment. Thai suction boats can hardly
he seen as a response to low grades – they worked deposits over twenty
times richer than those of the Malaysian gravel pump miners, but they
were certainly an interesting innovation which dramatically altered the balance of economic power in the industry.

In Brazil, as the industry started to gain pace towards the end of the
1970s, one might suppose that it would have been possible to transfer best
practice technology from elsewhere. Malaysia would have been an important
source, since it is widely acknowledged to have the world's most cost-effi-
cient alluvial tin industry. In fact, Brazilian deposits are somewhat different
from those in South-East Asia. Although there are some extremely rich
deposits, they often are quite small, and even on a single mining property
they may be too spread out to be mined with a single dredge. Thus,
although Malaysian mining consultancy companies were active in Brazil in
the 1970s, dredging technology has been greatly modified to suit Brazilian
conditions. Bucket ladder dredges used in Brazil have been of very small
capacity, with buckets of around 3 to 6 cubic feet.[16]

THE ECONOMIC STRUCTURE OF THE TIN INDUSTRY

William Fox, the late secretary of the International Tin Council, has argued
in his influential book on tin that the industry has had much of its form of
organisation determined by geology (Fox, 1974, p. 7). Tin is not only
scarce, in the sense of representing a small proportion of all minerals in the
earth's surface, but tends to occur in much smaller deposits than is the case
for other major non-ferrous metals. This has provided a niche for small firms
in the industry. Competition between small local producers and foreign-
owned firms has depended in the past, especially in Malaysia and Thailand,
on the development of efficient small scale mining techniques, which in
principle could also be scaled up to mine larger properties. While the smallness of most tin deposits may provide a niche for small operators, the
economic structure of the tin industry in relation to that of other minerals
depends more widely on factors such as differences in economies of scale in
production and processing, in the complexity of technology, and the importance of captive markets overseas as an outlet for raw material exports
(Vernon, 1973, p. 35).

Probably the most important dimension of the industry's economic structure for purposes of this book lies in the role and market power of foreign
mining companies in the developing countries who are the main producers,

and the relationships of these companies with the locally-owned sectors and state-owned enterprises. Thus, especially before the structural changes in world mining which started in the 1960s (Bomsell *et al.*, 1990), highly vertically integrated multi-national mining corporations have been able to exert great market power both in bargaining with host governments and in excluding local companies. Among the (non-oil) minerals, this can be illustrated by the case of bauxite, which has often been seen as an extreme example of MNC market power. Bauxite reserves are found quite widely, so MNCs can switch their source of supply if particular host governments are recalcitrant. At the same time the initial capital costs and high technology of extraction, and more particularly of refining and smelting, made entry difficult for private local firms, while state companies found it hard to sell their output (whether of bauxite, alumina or aluminium) in a world market dominated by the oligopolistic aluminium companies. The major changes since the late 1960s have affected aluminium less than many other metals, and the market power of the major multinationals in aluminium is still considerable (Auty, 1987, pp. 204–6).

The existence of successful small-scale technology in tin, though helped by the availability of many deposits too small for the methods favoured by foreign companies, also owes much (as later chapters will show) to successful experimentation by local miners. Underground mining in Bolivia owed much of its development, prior to nationalisation, to the Patino organisation, a locally originating multinational. In tin dredging the technology is not in itself too complex or too difficult to obtain. As the experience of local Asian dredging companies formed in Malaysia and Thailand in the 1960s has shown, management could be hired on the open market, though at that time it was often still expatriate. Nor is design technology monopolised by the mining multinationals. Indeed, it is more usually bought from specialist consultants. Further, there are in Malaysia engineering firms capable of fabricating and erecting dredges, who have also worked in Thailand, and who are predominantly local Chinese (like the miners). Both the existence of the design consultants and of local Asian engineering firms dates to before the second world war (Thoburn, 1973b). Since the 1985 tin crash, however, such engineering firms have tended to switch to other activities as tin business has declined.

The vertical structure of the industry also has not been a barrier to local mining companies entering tin. Taking the metal's main use until very recently (in packaging) there are four stages beyond mining, namely smelting, marketing, tinplate manufacture and can manufacture. MNC control over the smelting and marketing of Bolivia's production in the past has weakened the bargaining power of that country's state tin corporation and influenced the government to take measures to establish local smelting. In the case of most other producers, smelting (of alluvial ores) is subject to large economies of scale, but the technology was not too complex to prevent one large smelter in Malaysia being formed by local (ethnic Chinese) interests in the late nineteenth century. The degree of vertical integration between mining companies and smelting companies, though significant (especially between the two world wars), has not been overwhelming, and independently-owned mines and smelters have not been disadvantaged greatly by their lack of vertical links. Similarly most marketing has been done on an arm's length basis (though the use of long-term contracts is increasing). Smelting is a small part of the value of output, especially for alluvial ores, and generally no further refining is necessary.

Also important is the fact that there are very few links between smelting and mining companies on the one hand, and tinplate manufacturers on the other. Tinplate is overwhelmingly produced by steel companies and tin is a minor part of total costs. Tinplate manufacture in a country depends more on there being a domestic steel industry than on domestic tin production. The tinplate manufacturing industry in the past has been quite concentrated, however. Although it remains so in individual developed market economies (in the UK for example, where there is only one producer, British Steel) tinplate manufacture has spread widely to developing countries. In 1960 62 per cent of world tinplate was produced by the US (and Canada). By 1970 this had fallen to 41 per cent and by 1982 to 22 per cent (*TI*, January 1986). Buyer concentration gave to individual manufacturers fearing for their supply sources and wishing to integrate backwards to secure them. In fact, the tendency for tin to be sold on the open market has greatly alleviated any potential supply insecurity. Moreover, historical processes have tended to cumulate, and the lack of backward integration by some companies has meant that the others have less incentive to move backwards. In solder, however, there have been some links between smelters and solder manufacturing. Both Malaysian smelters manufacture solder (Meyanathan, 1988), and Billiton, the owner of the Thaisarco smelter in Thailand, extended its range of solder manufacture in the 1970s (*TI*, September 1978).

In tin can manufacture, according to a study in the late 1970s by the International Metalworkers Federation (reported in *TI*, September 1978), five MNCs dominated the world market. These were Metal Box of the UK and four large US companies: the Continental Group, American Can Company, National Can and Crown Cork and Seal. Their position had been greatly strengthened, apparently, by the development of two-piece can manufacture, with its much higher investment requirements than those of the more traditional three-piece can. This industry structure could have made for difficulties for producers moving into downstream activities, at least if they wanted to compete on world markets. In Malaysia, where domestic consumption of tin cans is met mainly by domestic production, Metal Box has been a major foreign investor, but there are also Malaysian and Singaporean companies involved (Meyanathan, 1988). In tin-mining, the structure of the world market for tin metal does not seem to be a factor enhancing the power of MNCs relative to local companies. More important a barrier to local entry in the past has been the high capital costs of dredging relative to the smaller scale techniques,in a situation where local Asian (and African) producers normally used partnerships or private companies rather than public limited companies. In Malaysia or Thailand prior to the 1985 tin collapse, a gravel pump mine could be established for about US$500,000, whereas a large inland dredge project would cost about US$15 million (Thoburn, 1981a, p. 115). One of the most advanced and the largest dredges ever built (Billiton's offshore dredge in Indonesia) was part of a project in the 1970s costing US$65 million, of which $30 million was for the dredge alone; in contrast, a Thai suction boat would have cost only about US$75,000. The only tin project above $100 million was said to be that to develop the Kuala Langat in Selangor state in Malaysia, which is worked as a deep multiple dredging operation. In Bolivia, an underground mine would have taken about US$13 million to establish (Engel and Allen 1979, p. 145). More recent investment figures are not available for South-East Asian and African producers, because of an almost total absence of new tin projects since the tin crash.[17]

It is important to emphasise, though, that by world mining standards these dredge and underground mining investment costs do not make tin mining a large-scale operation. According to Bosson and Varon's (1977, p. 46) study, based on 1973 data, 75 per cent of the total investment expenditure in a large sample of mining ventures was in projects costing over US$100 million, and inflation later in the 1970s would have increased those costs. In other words, capital costs in tin have been high enough to deter local firms in the past from investing in certain mining techniques, at least in South-East Asia and in Africa, but the really large investments needed in other metals have not been in evidence. In consequence the services of large international mining houses have not been required to develop the industry in the past (Fox, 1974, p. 8). Instead, tin companies historically have been specialist, exemplified by the London Tin Corporation, which controlled about a quarter of world tin mine production at the peak of its power in the 1930s, and still about an eighth in the 1970s prior to its takeover by the Malaysian government (Fox, 1974, p. 9). The fact that most tin is mined by alluvial methods, whereas dry, open-cast mining is increasingly used in other metals, has also been a factor making for some separateness in terms of technical expertise.

The lower investment costs in tin compared to other metals reflects the fact that even the largest tin operations are relatively small in terms of their physical output. Before the exploitation of the huge Patinga and Bom Futuro deposits in Brazil in the 1980s, 3000 tons a year would have been an exceptionally large output for a tin mining property.[18] In contrast, copper mines could have been operating at 60,000 tons a year or more, and lead and zinc mines at well over 30,000 tons (Robertson, 1982, p. 1). At 1990 prices (see Table 1.6) such operations would have yielded substantially greater output values than for a tin mine. In terms of total world output value too, tin is of low importance compared to many other major metals; its 1990–1 output value was 5 per cent or less of that of aluminium, copper or gold, 11 per cent of that of zinc, 17 per cent of that of nickel, and 37 per cent of that of lead (Crowson, 1992b, p. xv). In 1990 it accounted for only 1.1 per cent of the value of all non-fuel mineral production in the Western world, the lowest of fourteen major metals (Roskill, 1992, p. 7). Of course, this low share is also due in part to the low relative price of tin in the 1990s.

The structural changes which started in other metals in the 1960s and the 1970s have affected tin to some extent too. State enterprise has grown. The largest three companies in Bolivia were nationalised in 1952. In the 1970s a Malaysian government agency took control of the London Tin Corporation through a stock-exchange takeover, eventually to form the Malaysia Mining Corporation, which now controls most of Malaysia's previously foreign dredging sector and one of the country's two smelters. Thailand has also had some growth of state enterprise in the form of the Offshore Mining Organisation. Indonesian tin had had a long history of state involvement under Dutch colonial rule, and the Indonesians took over the remaining private Dutch interests in 1957–8. Most of the industry is now controlled by the state enterprise P. T. Timah. In the 1960s in Zaire and the 1970s in Nigeria, state involvement in the industry also grew. Brazilian tin mining has been developed predominantly by indigenously-owned companies, with little involvement of MNCs or the state. Bolivia is in process of privatising some of its state mining operations. There has also been some entry into tin by multinational mining companies diversifying from other metals, and Shell and BP among the oil companies bought up some tin interests. Since 1949, the industry in China has been run by the state.

TIN MINING AND ECONOMIC DEVELOPMENT

In the 1960s and 1970s there was strong interest among LDCs generally in increasing their gains from primary commodity exporting. This reached a peak in the demands for a New International Economic Order made through the United Nations in 1974, following the cartelisation of the oil industry by the Organisation of Petroleum Exporting Countries (OPEC) in 1973–4. The agenda for primary commodities stressed the need to raise the level of commodity prices and to increase the stability of export earnings. Eighteen primary commodities were identified as being of interest to LDCs, of which ten were regarded as especially important and suitable for market regulation. These ten were the so-called 'core commodities': cocoa, coffee, cotton, hard fibres, jute, rubber, sugar, tea, copper and tin (Thoburn, 1977b, ch. 2). It was proposed through the United Nations Conference on Trade and Development (UNCTAD) that a Common Fund should be established to coordinate and support then-existing commodity agreements (tin, coffee and sugar) as part of an Integrated Programme for Commodities, and to encourage the formation of new ICAs. Tin, with its predominantly Third World sources of supply and already existing ICA, was an archetypal UNCTAD core commodity. In the event, the only new ICA stimulated by the UNCTAD initiative was rubber, and the Common Fund was not established as an organisation until 1989 (Maizels, 1992, chs 7 and 8).

There was also a belief that gains could be increased through the 'restructuring' of primary export production. In many minerals this involved governments adopting a more aggressive policy towards foreign investors, and sometimes promoting state mineral enterprises as an alternative. The 1960s and the early 1970s were years when mineral prices were high in relation to costs, so there were significant gains to be bargained over. This was particularly so in tin for the intramarginal producers, where high prices continued through the late 1970s.

In principle, policies to increase host countries' development gains from mineral exports can include the ownership restructuring of foreign firms, taxation, measures to promote forward and backward integration, and the encouragement of wider development effects such as labour training (Thoburn, 1981b, pp. 75–80). In tin, ownership restructuring has included the nationalisations and stock exchange takeovers mentioned in the previous section. There have also been various restrictions imposed in most producing countries on the proportion of equity which can be held by foreigners. Taxation policy in mining has been much discussed by economists because it is often argued that mining, especially in LDCs, has so few positive effects on development that the taxation of mining revenue is the main way in which the host economy can secure some economic gain. The scope for taxation is greater in mining ventures if rich deposits are mined simultaneously with poorer ones, as in tin, such that the former generate returns in the form of resource rent (or 'mineral rent'), in the sense of profit over and above the supply price of investment. The monopolistic and oligopolistic structure of mining industries gives rise in addition to monopoly rents,[19] which also can be taxed, and the incidence of taxes varies with market structure. Tax policy can, of course, be combined with provisions to strengthen other development effects so as to reduce further the enclave character of mining. Policies such as equity participation, management contracts, and production sharing are alternatives to tax policy (Thoburn, 1981a, p. 21).

Tax policy is also relevant to state enterprises, since it cannot be assumed that profits retained by such enterprises will necessarily be used in the interests of national development (Gillis *et al.*, 1978, p. 10). In several of the major tin-producing countries, export taxation on tin dates back to long before the second world war. The combination of export taxation and taxes on profits became highly contentious in the 1970s, as foreign investors argued that taxes were 'penal', particularly in the largest producer, Malaysia (Thoburn, 1978b).[20]

Policies on backward and forward integration in tin, and on tin's wider development effects, have to be seen in relation to the effects of tin exports on economic development which have occurred through market forces and through policy in the more distant past. The extensive literature, established by the 1970s, on the effects of primary commodity exports on economic development, shows that in a given situation different export commodities are likely to have different development effects (Baldwin, 1963; Watkins, 1963; Hirschman, 1977).[21]

These development gains can be measured in the first instance by estimating *retained value* (RV). This is the share of a given export income which accrues to local income recipients. It may reflect local employment, backward integration into the local production of intermediate products or capital goods, or local taxation and other means (such as equity sharing) by which the share of foreign profits and other foreign factor incomes has been reduced. In empirical work on trade and development it has been especially important to determine the distribution of gains from exporting, first between local and foreign income recipients, and then (within the latter category) between workers, profit receivers and the government. In consequence, the calculation of retained value is a natural starting point, and has become a common tool in commodity studies generally.[22] Its use as a measure of real income gain requires qualification, though, and this is discussed when the measure is used in Chapter 5. Also important is the host country's share in the export good's *final consumer value* (FCV). This involves forward integration into processing or the manufacture of final products, or increasing local participation in marketing and distribution.

Estimates of RV and FCV also generate data with which to analyse a mineral export's development effects more generally. An export opportunity, perhaps initiated by foreign investment, may stimulate domestic firms to enter the industry. Opportunities to produce intermediate products or capital goods for the export sector, or to use the export good as an input for further processing, likewise may stimulate investment. These opportunities are, of course, *backward and forward linkages* in Hirschman's (1958) terminology. There may also be final demand or 'consumption' linkage (the creation of local markets for consumer goods by the expenditures of export sector workers) and fiscal linkage (i.e. tax revenue paid to the government) (Hirschman, 1977). If the country is short of investment outlets, linkages may raise the overall level of net investment and stimulate new saving. Even if the linkages are simply reallocations of existing investment resources, new technology may be introduced with them which raises the productivity of local capital and labour. In addition, workers may improve their skills through these new employments, and managers and entrepreneurs may learn new and useful methods. These improvements may, in turn, increase the responsiveness of the domestic economy to linkage and other new investment opportunities, for which export sector profits may help provide capital. Linkage investment too (e.g. in transport) may stimulate growth in other sectors.[23]

In terms of these measures of development effects, however, mining does not necessarily have a favourable impact on the economy of an LDC. It is often stylised as an activity where technology is so complex and the capital investment requirements so large that only foreign investors can operate. Since such mining is highly capital-intensive, value-added will consist mainly of profits, which, in the absence of taxation, will accrue to the foreign owners and be remitted overseas. The wage bill is unlikely to be large enough for wage-earners' purchases to generate a market for local consumer goods, and thereby spread development, and such workers may be expatriate anyway and consume imported goods. Few local intermediate products or capital goods will be purchased locally, because an LDC would not have the industrial skills to produce them, and the output may be processed abroad. Mining of this sort conforms to the stereotyped *export enclave* of the structuralist literature on development (e.g. Singer, 1950), where LDCs lose to the industrial countries the benefits of their primary commodity exporting (Thoburn, 1977b, ch. 3). Such views have made Lanning and Mueller (1979, p. 24) ask in their study of Africa: 'when the companies have exhausted Africa's major mineral deposits, will the continent have anything more to show than huge holes in the ground?'

Many minerals do appear to conform to this stereotype, and in such cases, as already noted, taxation is the main way LDCs may be able to secure development gains from them. Tin, however, is a rather exceptional mineral in this regard, and has had substantial and positive development effects in several producing countries, which will be discussed in Chapter 5. These effects have worked through local purchases of intermediate goods, local market creation, the local processing of tin ore into smelted tin, and labour training, as well the generation of substantial tax revenue. These favourable effects have characterised even the foreign-owned sectors, and in the smaller-scale, locally-owned activities they have been even more pronounced. The development of local tin smelting in major producers has to some extent improved the bargaining position of tin exporters in relation to foreign buyers, since metal can in principle be sold direct to final consumers. Hughes and Singh (1978) for instance showed that considerable economic rent accrued in many minerals mainly at the processing stage, and has not been appropriated by LDCs who continue to export the unprocessed product. They concluded that tin-exporting countries in the mid-1970s had succeeded in securing the largest share of the economic rent, in relation to consuming countries, among the five major minerals they studied (the others were petroleum, iron ore, phosphate rock and bauxite).

Another interesting aspect has been the competition historically between local mining companies and foreign firms, which has expressed itself in South-East Asia through the competitive development of mining technologies. It is also important to know that in South-East Asian tin mining the distribution of gains has had strong ethnic overtones. Tin mining in Malaysia and Thailand has been as industry where there has always been a substantial local presence, but this has mainly been in the form of mines owned by ethnic Chinese immigrants, not indigenous Malays or Thais. In Indonesia, despite a dominant state and foreign company presence, immigrant Chinese had a substantial, and pioneering, involvement too as workers and producers. The industry in Malaysia and Thailand was expanded by immigrant Chinese in the late nineteenth century using open-cast, dry mining methods. In both Malaysia and Indonesia, though, some tin mining had been done previously by local people. By the first world war, gravel pumping and dredging were establishing themselves as the main mining methods, with the former used mainly by ethnic Chinese and the latter by Western companies. Ethnic Chinese workers have always constituted a large proportion of the mining workforce in South-East Asia.

All of the development effects of mining in the Third World mentioned so far apply even if mining activity in the LDCs concerned constitutes a small proportion of national output and exports. In some economies, however, mineral exports occupy a more important position. Nankani (1979, p. i) has labelled countries as 'mineral export economies' where minerals generate more than 40 per cent of exports and 10 per cent of national output. Although any such choice of figures is arbitrary, the distinction is to highlight the fact that where mineral exports are so important, the growth of such exports has far-reaching effects on development. Such effects work through key prices, particularly the real exchange rate and the level of wages. They also influence income distribution, and with it the structure of demand and the pattern of savings. Appreciation of the real exchange rate resulting from booming mineral exports can damage sales of the country's other export products and it can increase import penetration in local agriculture and manufacturing. This is the so-called 'Dutch disease' problem, named after the effects of North Sea gas exports on the Netherlands economy. Where mineral rents generate a large proportion of the government's revenue, the use to which such revenue is put is a crucial influence on the country's development. There is great scope for policy in managing the mining sector for development, for good or ill (Auty, 1991a, 1991b; Daniels 1990). As Tilton (1992, pp. 1–2) has observed, the development performance of many mineral export economies in recent years has been extremely disappointing, and their poor record cannot simply be attributed to the low mineral prices which have prevailed since the mid-1970s.

Successful mineral exporters in the long term diversify beyond their mineral export base and cease to be mineral economies. Among the producers of tin there are some 'mineral economies'. In recent years only Bolivia has been such a one based on tin, and even in this case the dependence is distorted by the exclusion from official statistics of the country's illegal exports of cocaine. In the nineteenth century Malaysia certainly was a mineral economy based on tin, and tin held a significant though lesser place in the early development of Thailand and to some extent Indonesia. Zaire is still described as a mineral economy, but not because of tin, and Indonesia's mineral export dependence is on oil. The uses of tin revenue in producers in the past is a theme to be taken up in later chapters.

INTRODUCTION TO THE MAIN TIN-EXPORTING COUNTRIES

This section provides brief profiles[24] of the main tin-exporting countries, basic economic data for which are set out in Table 1.9. All six countries, except Bolivia, have experienced high rates of growth in the last twenty-five years, although the long-term figures obscure the acceleration of growth in China in the 1980s and the slowing down of Brazil's. The three South-East Asian economies have all undergone great structural change, particularly towards a greater export orientation. All the major tin exporters, except Bolivia, have greatly expanded their manufactured exports, particularly in relation to agricultural exports. In the post-war period tin has been far more of an important export earner for Malaysia and Bolivia than for the other countries. Now, tin is not a major export for any of them, except to some extent for Bolivia. Bolivia's tin exports may recover if the rehabilitation of the industry succeeds.

TABLE 1.9   Economic profiles of major tin exporting countries, 1990

	Malaysia	Indonesia	Thailand	Brazil	Bolivia	China
Population (millions)	17.9	178.2	55.8	150.4	7.2	1133.7
Land area (000 km²)	330	1905	513	8512	1099	9561
Gross national product per head ($)	2320	570	1420	2680	630	370
(Gross domestic product per head at purchasing power parity exchange rates ($))	(5900)	(2350)	(4610)	(4780)	(1910)	(1950)
Average annual GNP growth rate 1965–90 (%)	4.0	4.5	4.4	3.3	-0.7	5.8
Average annual inflation growth rate 1980–90 (%)	1.6	8.4	3.4	284.3	317.9	5.8
Life expectancy at birth (years)	70	62	66	66	60	70
Percentage shares (1989) in GDP of:						
agriculture	19	22	12	10	24	27
manufacturing	27	20	26	26	13	38
mining and quarrying	10	13	3	1	14	NA
Percentage share of exports in GDP	79[42]	26[5]	38[16]	7[8]	21[21]	18[4]
Percentage share in merchandise exports of:						
fuels, minerals and metals	19[34]	48[43]	2[11]	16[9]	69[93]	10[15]
other primary commodities	37[60]	16[53]	34[86]	31[83]	27[3]	16[20]
manufactures	44[6]	35[4]	64[3]	53[9]	5[4]	73[65]
Tin as percentage of total merchandise exports:						
1990	1.13	0.68	0.32	0.58	11.7	0.19
1978	11	2	7	NA	57	0.45
1965	28	6	6	–	79	1.9
Debt service ratio (debt service as % of exports)	11.7	30.9	17.2	20.8	39.8	10.3

Sources and Notes
1. Sectoral shares of GDP for Malaysian agriculture and manufacturing are estimates from Malaysia (1991); sectoral shares of GDP for mining and quarrying are from UN (1991c); tin export figures for 1990 are from UNCTAD (1992a), and 1978 from Thoburn (1981a, p. 46) for Malaysia, Indonesia and Thailand; 1965 tin export figures are from UN (1967b and 1970b).
2. All other statistics are from World Bank (1992).
3. Purchasing power parity GDP figures are expressed in 'international dollars' to give the same purchasing power over GDP as a US dollar in the USA (World Bank, 1992, pp. 299–301)
4. Share of tin exports in total exports for China for 1965 and 1978 is based on Chinese tin export figures from Williamson (1984), valued at average annual tin price; Chinese total export figures are from SSB (1991).
5. Figures for 1965 are in square brackets.

Malaysia

Malaysian growth in the past has been closely associated with primary commodity exports. It is an unusually open economy and has been so since the great expansion of tin exports in the late nineteenth century and the growth of the rubber industry in the early years of the twentieth. Rapid growth in oil palm and timber exports in the post-war period lessened dependence on the traditional export staples. As of 1990 Malaysia was still the world's largest exporter of natural rubber, just ahead of Indonesia, and by far the largest exporter of palm oil (UNCTAD, 1992a). The country was under the control of the British from 1874 to 1957, and British firms have accounted for the bulk of foreign ownership of the mines and plantations. Chinese immigrated into Malaya in large numbers in the nineteenth and early twentieth centuries, and now comprise about a third of the total population, owning much of the smaller-scale mining sector and being important in trade, small industry and small rubber estates. Indian workers were imported to work on the large rubber estates and they form about 10 per cent of the total population. The indigenous Malays, or *bumiputras* (a term which also includes other indigenous people such as Dayaks, who live mainly in East Malaysia [Sabah and Sarawak on the island of Borneo]), traditionally have formed the bulk of the rural population, among whom rice growing was the predominant activity, but Malays also own well over half of the acreage of rubber smallholdings (defined as holdings under 100 acres).

The highly multiracial nature of Malaysian society has led to some distinctive problems. Serious racial riots in 1969 broke a generally accepted implicit understanding that Malays would hold the political power and Chinese the economic power. As a direct result, a New Economic Policy (NEP) was set up, to run from 1970 to 1990. This policy aimed at the 'restructuring' of asset ownership and employment in favour of the Malays, though it also had the declared aim of eradicating poverty regardless of race. The NEP has been succeeded by a National Development Policy (NDP), which also stresses restructuring, but (unlike the NEP) does not have quantitative targets or time limits. The Malaysian government, under its Vision 2020 idea, hopes Malaysia will become a developed country by early in the next century.

Manufacturing, first introduced as part of an import substitution policy in the early years after independence, has played an increasing role in Malaysian development. Large amounts of foreign investment have been attracted into export manufacturing, particularly electronics and textiles. Manufactures by 1990 were generating nearly half of total exports.

Malaysia is also a producer of oil, and it has one of the highest per capita incomes in Asia. Over the long term it has sustained a high rate of economic growth (see Table 1.9). It suffered severely from the general commodities price collapse in the 1980s, but has resumed growth helped by large inflows of direct foreign investment in manufacturing (World Bank, 1989).

Tin is mined primarily in the two states of Perak and Selangor on the west coast, which are also the centre of economic activity more generally. Tin and petroleum are Malaysia's main minerals. There is also some mining of copper, bauxite and iron ore.

Indonesia

Indonesia achieved independence from the Dutch after the second world war and occupation by the Japanese. Now that the Soviet Union has disintegrated, Indonesia is in terms of population the fourth largest country in the world. It combines exceptionally high population density in the main island of Java (where nearly 60 per cent of the population live) with considerable unused land in its numerous outer islands. After independence, under President Sukarno's 'guided economy' the country reached a peak of dislocation in the mid-1960s, accumulating high external debt and experiencing an inflation rate which reached 650 per cent. Many sectors were run down and per capita income growth was reversed. The new military regime which took over in 1966 placed a heavy stress on stabilisation, and has been in power ever since. Foreign investment was encouraged too, under the Foreign Investment Law of 1967, the Dutch having been expelled by Sukarno in the late 1950s from their main economic activities (including rubber estates and the tin mines).

Like Malaysia, Indonesia has had a long history of exporting tin, rubber and palm oil. In 1990 it was the world's second largest exporter of natural rubber, close to Malaysia (UNCTAD, 1992a). It also has a history as a petroleum exporter, with petroleum production dating back to before the first world war. As a result of the OPEC price rises of the 1970s, by the end of the 1970s petroleum alone was generating more than half of export earnings and of government revenue. The oil price collapse in the mid-1980s seriously affected Indonesian growth, and Indonesia started a successful programme of structural adjustment, including devaluations and liberalization of the trade regime (Tambunan, 1990). By the early 1990s the share of oil in total export earnings had dropped to about a third. Like Malaysia and Thailand, Indonesia has attracted foreign investment in export-orientated manufacturing (*FEER*, 22 April 1993).

The supply of food and the creation of employment probably have been the country's two most serious problems. From being the world's largest importer of rice in the 1970s, Indonesia became self-sufficient in rice in 1984 (*The Economist*, 17 April 1993). Employment creation has been a problem common to other countries discussed here, though Indonesia's population growth rate has fallen somewhat, as the result of a successful birth control programme. It has the added dimension, however, of involving considerable resettlement of people, and there has been a transmigration programme to settle families from the heavily-populated central areas to the outer islands.

New minerals have risen rapidly since the late 1960s. Most minerals are mined outside of the main areas of population, copper, nickel and bauxite being the most important ones besides tin. Almost all tin is obtained from the islands of Bangka, Belitung and Singkep, off the coast of Sumatra. Chinese immigration, as in Malaysia and Thailand, was associated with the tin industry. The Chinese today number about four million, but they are important in business and commerce. Many were massacred as part of the military takeover in 1966 when they were associated with an unsuccessful communist attempt to seize power. Like Malaysia with its policy towards bumiputras, Indonesia has had a declared aim of restructuring its economy in favour of indigenous Indonesians (*pribumis*).

Thailand

Thailand is unusual among developing countries in never having been a western colony. It formed instead a buffer between British Malaya to the south and French Indochina to the north. The process of export expansion and colonial rule in Malaya was a powerful influence on the Thai economy nevertheless. Free trade was imposed on Thailand by the Bowring treaty of 1856 with Britain, and rice exports grew rapidly to a third of the world total before the second world war, Malaya being a major customer. Other exports grew too, particularly rubber, tin and teak, but by the late 1920s rice was still about two-thirds of the total. During the late nineteenth century and for most of the interwar period tin formed about 10–15 per cent of its exports (Ingram, 1971, p. 94). Tin thus never occupied the dominant position it held in Malaya before the coming of rubber although, as in Malaya, it was an important factor in inducing Chinese immigration. The Chinese now form some 14 per cent of the population, though they are assimilated to a much greater degree in Thai society than has occurred in Malaysia or Indonesia.

Thailand is a constitutional monarchy, royal power having been limited by a military coup in 1932. The military in practice were the dominant element in the government for most of the time until the 1980s, and entered politics directly again in the early 1990s.

The dominant position of domestic and export agriculture continued well after the second world war. Since the 1960s, Thailand has achieved a high rate of growth. This was based initially on further increases in agricultural output, mainly in the form of an expansion in the cultivated area.

Thailand is still the world's largest exporter of rice (UNCTAD, 1992a). Apart from a brief and unsuccessful period of direct government investment in industry in the 1950s, industrialisation began in the 1960s with an import substitution programme. Foreign investment has been officially encouraged since the 1960s, but of the capital of the firms granted promotion certificates over the 1959–69 period, about two-thirds were Thai and much of the foreign capital was in majority Thai-owned joint ventures (Ingram, 1971, pp. 290–92). Manufacturing development began to be more export-orientated in the 1970s aided by a *de facto* depreciation of the Thai baht which drifted downwards with the US dollar to which it was tied until 1978. In the 1980s and 1990s Thailand has been a major recipient of foreign investment in export-orientated manufacturing (less so than Malaysia, but more than Indonesia) and in the late 1980s was sustaining a double-digit rate of economic growth. Much of its development has been centred on greater Bangkok, and there are severe infrastructural and environmental problems.

The move towards export manufacturing has been more employment creating than was the import substitution programme. This, combined with the agricultural growth, has meant a quite widespread participation among the population in the growth of income. Thus the World Bank (1980) estimated that the proportion of people in absolute poverty has fallen from a half to a quarter since during the 1960s and 1970s. Nevertheless many farm households, especially in the north and north-east, have not gained greatly, and this continues to be a problem.

Mining in Thailand is less important than in most of the other tin producers, and it is dominated by tin. Most tin comes from the southern provinces, particularly Takuapa, Phanggna, Ranong and Phuket. This concentration is largely the result of geology, but historically mining has been restricted to the south by government decree

Bolivia

In the middle ages Bolivia was an important part of the Inca empire, and was conquered subsequently by the Spanish, for whom it was a major supplier of silver. Mining, first of silver, then of tin, and more recently also of petroleum and natural gas, has been a central feature of Bolivian development. The country has been independent since 1825, during which period it has had an unusually large number of changes of government. The important change of government in 1952, when the nationalist revolutionary movement took power, was closely associated with the power of the tin mine workers. At that time the export of minerals, principally tin, generated 97 per cent of the country's export earnings, over two thirds of which was produced by the three large mining companies which were nationalised to form the state mining corporation Comibol. James Dunkerley (1984) has suggested that Bolivian society is a system which is able to 'manifest continuity and logic precisely in instability, and frequent, if not constant, shifts in the locus of power'. The main countervailing powers historically have been the armed forces and the labour movement based on the tin mine workers. From 1964 to 1982 the country was ruled by a series of military governments, including the particularly harsh dictatorship of General Luis Garcia Meza (1980–2), whose troops were responsible for a notorious massacre of mineworkers in an attempt to crush their union.

Bolivia has been landlocked since the late nineteenth century when it lost its coastal region in a war with Chile. It also lost territory in the southeast following war with Paraguay in the 1930s. Communications in the past have tended to run from the mountains to the Chilean coast, for the benefit of the export of metals. Bolivia has a land area more than half that of Indonesia, yet has only 4 per cent of Indonesia's population. Compared to Malaysia and Thailand too, Bolivia is very sparsely populated. The main area of population (about three-quarters of the total) is the high plateau area or *altiplano* and the eastern cordillera of the Andes. From this mountainous area a series of high valleys lead down to the low-lying eastern region. Most of the tin and other metals is found in the mountainous areas in and around the altiplano, and the great height of the mines as well as their workings being underground adds to costs. The climate of Bolivia varies from subtropical in the lowlands to the extreme cold of the mountains in winter. Oil is found in the lowlands, where there is also considerable potential for agricultural development, though this has been hampered in the past by lack of infrastructure. Roughly half of Bolivia's population is made up of indigenous Indians, the highest proportion in Latin America, who historically have been disadvantaged in a variety of ways.

In the 1980s Bolivia suffered from severe external debt problems and the ending of its petroleum export surplus. Hyperinflation had developed by the mid-1980s. To deal with these difficulties there have been extensive programmes of structural adjustment for the whole economy, with World Bank and International Monetary Fund involvement, and by the start of the 1990s the annual rate of inflation was down to under 10 per cent.

Comibol was in a state of long-term decline, with rising costs, declining grades and lack of long-term investment well before the 1985 tin price collapse (Jordan and Warhurst, 1992). The adjustment measures have involved many closures of mines run by Comibol. Most recently, privatisation of Comibol mines, along with joint ventures with foreign companies, has set off considerable opposition among tin miners with their long tradition of union militancy.

Comibol is also involved in the production of other minerals, principally zinc, lead and silver. In 1990 metallic minerals contributed some 44 per cent of total exports, of which tin was roughly a quarter (*Mining Journal*, 1991, p. 52). Bolivia also has a large illegal economy based on the export of cocaine, of which the country is said to generate over 40 per cent of world supply (Crabtree *et al.*, 1987, p. 21). There is drugs trade between Brazil and Bolivia, with Brazilian tin being exchanged for cocaine. Drug money too is an influence in the political arena. It has been argued that it was the Bolivian military's failure to take full control of the cocaine industry which led to the return to democratic rule in 1982, with the military withdrawing to maintain its internal cohesion (Dunkerley, 1984).

Brazil

Brazil declared its independence from Portugal in 1822. Under Portuguese rule it had been an important exporter of sugar and then of coffee, which became its main export in the nineteenth century. Industrialisation started under military rule in the 1930s, and a period of accelerated industrialisation under another military government from 1964 laid the basis for the ten years of the 'Brazilian miracle' (MacDonald, 1991; Hewitt, 1992). Over the long period 1940–80, Brazil averaged the very high per capita income growth rate of 4 per cent per year. The oil price rises of 1973–4 and 1979 caused severe problems for Brazil, and were met initially by a build-up of external debt. Severe structural problems became apparent in the 1980s, with a large budget deficit and an acceleration of inflation. National income fell during the late 1980s, and the consumer price index rose from 100 in 1986 to 3 million in 1991 (*The Economist*, 7 December 1991). Brazil has one of the most unequal income distributions in the world. The top 10 per cent of households have the highest share of income (46.2 per cent), and the bottom 20 per cent of households the lowest share (2.4 per cent), of any country for which data is listed in the World Bank's *World Development Report* (1992).

When tin in Brazil was first attracting attention in the 1970s, the country was being seen more generally as a source of new material discoveries, and the Brazilian government was attaching much importance to the development of new mineral production (Lloyd and Wheeler, 1977). In the 1980s there was a $13 billion ten-year development plan for the non-ferrous minerals sector, focusing on the Amazon region (Crabtree *et al.*, 1987, p. 38), now the centre of tin mining. Now, Brazil is the world's tenth largest exporter of minerals, and larger than any LDC except Chile. As a newly industrialising country, it is also a major importer of minerals and metals (twentieth in the world), 'though the value of its exports are three times larger than its imports. Brazil is a medium-sized producer of oil, with output in 1990 slightly more than that of Malaysia, and about a third that of the UK; but this is sufficient for only about half of domestic needs

(UNCTAD, 1992a). Brazil is the world's largest exporter of iron ore, and the third largest of manganese (UNCTAD, 1992a), as well as having the world's third biggest reserves of bauxite (Crowson, 1992b, p. 1). Compared to Brazil's exports of iron ore and bauxite/aluminium, its tin exports are of small importance in its mineral export earnings (DNPM, 1990, p. 54).

Foreign investment has been encouraged though on fairly stiff terms and normally in joint ventures. Companies involved have included British Petroleum, Bethlehem Steel (in manganese), Alcan (bauxite), Metallgesellschaft (nickel) and Mitsui and Sumitomo (chrome) (Lloyd and Wheeler, 1977). Investment in minerals is an area restricted in relation to other sectors (Dunning and Cantwell, 1987, p. 658). In spite of Brazil's large geographical area, most of the population live on the narrow coastal strip and most mineral discoveries have been further inland. Tin production first started in 1903, but there was no production on a commercial scale until the mid-1940s (Engel and Allen, 1979, p. 63), and Brazil was not a significant producer until the 1980s. Unlike other producers, tin smelting was developed in Brazil ahead of domestic mine production, and Brazil imported tin concentrates to smelt for its domestic tinplate industry.

China

As noted earlier, China has been producing tin for at least 2000 years, but the overall importance of tin in such a large economy has always been negligible. Therefore the main interest in China in this book is in the effect of Chinese production on the world market. Tin in China is found in the south, particularly the south-west province of Yunnan and the southern province of Guangxi, which together have half of China's tin reserves (*MT*, 15 April 1986). There are also tin deposits in Guangdong, Hainan, Hunan and Jiangxi provinces.

Exports of tin from China since the communist revolution of 1949 have fluctuated considerably, but increased substantially in the late 1980s (UNCTAD, 1992a). Major reforms of the Chinese economic system were initiated in 1978, after Chairman Mao Zedong's death two years previously. From 1949 until the reforms, exports were determined on the basis of import requirements, and exporting enterprises received prices in domestic currency which bore little or no relation to international prices. Since 1978, and more especially since the mid 1980s, Chinese domestic export prices have been brought more in line with world prices, and there has been a series of devaluations of the Chinese currency (the *Renminbi*), which have raised prices to exporters (Lardy, 1992). From 1949 until the start of the reforms there was no Western direct investment, and even before the revolution, there was strong resistance to any Western involvement in tin mining (Jones, 1925, p. 240). Reports on China in the 1980s indicated some moves towards joint ventures in tin to develop new deposits (*MT*, 15 April 1986)

SUMMARY

Tin is a commodity for which Western countries and Japan are unusually dependent on Third World sources of supply. Tin had the longest-lived and apparently most successful international commodity agreement, which operated continuously from 1956 until 1985, when it collapsed in spectacular default. Tin had suffered from a long-term slowdown in consumption, due in part to technical change which economised on the tin content of tin-plate and solder, its main uses, and in part due to competition from other materials, particularly aluminium. In contrast to aluminium, whose real price was falling until the 1960s, tin had experienced a constant real price trend during the first half of the twentieth century. The rising price of tin in the 1960s and 1970s, in the face of world industrial growth and apparently scarce tin resources, hastened tin's loss of competitive position. The world recession of the early 1980s coincided with the rise of a major new producer, Brazil, whose search for tin reserves had been encouraged by the high prices. Tin, like most other primary commodities, experienced historically low prices in the 1980s, but in more acute form. The International Tin Agreement proved unable to hold off from the market the huge stocks which accumulated as demand fell and sales from Brazil (and China) rose.

The United States has virtually no tin deposits of its own. It has maintained throughout the post-war period a large strategic stockpile of tin, disposals from which have been a worry to tin-producing countries. In part, American stockpiling, as later chapters will show, reflects the US's bad experiences in the face of the power of a British-backed tin cartel in the inter-war period.

Within several of the main producing countries, tin also has been unusual, at least among other metals, in having a large and viable small-scale sector. This has given local miners a chance to participate in the gains from tin exporting. In South-East Asia these local miners have predominantly been of ethnic Chinese origin, and were responsible for most of the early development of the industry before Western foreign investors became established. The existence of these small-scale operations is partly due to the nature of tin deposits, which are smaller than those for many metals. They also result from successful technical experimentation by effective local entrepreneurs, who developed the mining technique of gravel pumping in competition with foreign firms. Even among the foreign firms in tin mining, the companies have tended to be specialist tin producers. The major mining multinationals who dominate other metals did not enter tin until the 1960s and 1970s, and then only to a limited extent. They mostly have exited following the post-1985 price collapse. One consequence of the structure of the industry has been that tin exporting, at least in South-East Asia, probably has had a more favourable effect on development than virtually any other major metal produced in the Third World.

Tin has seen nationalisations in Indonesia and Bolivia and government-induced restructuring of foreign companies' equity in Malaysia and Thailand. Indonesia has a long history of state involvement in tin, going back to Dutch colonial times. The high tin prices of the 1960s and 1970s generated large mineral rents for intramarginal producers, particularly those in South-East Asia, and the producer governments who tax them. The Bolivian tin industry, which mines underground deposits, has had a long history of high costs and other problems, which are by no means yet solved.

Since the 1985 price collapse, tin demand has revived somewhat as industrializing LDCs increasingly have become users of tin. Tin's market prospects also have been helped by the growth of the world's electronics industry, and by environmental concerns which have led to demands to reduce the lead content, and increase the tin, in various tin-using products. However, the collapse of the Soviet Union has had a serious, though possibly temporary, effect on tin prices, since the USSR previously was a large importer. Most producers, except (to some extent) Brazil, have been hard hit by the continuing price collapse, and their tin mining has greatly contracted, especially in the small-scale sector.

NOTES

1. This point is well made by Rees (1990, pp. 85–104).
2. For a discussion of the tin industry's cartelisation potential prior to the collapse of the International Tin Agreement in 1985, see Thoburn (1982a).
3. For an analysis of the development effects of tin exports in Malaysia, the world's largest producer for most of this century, see Thoburn (1973a, 1973b and 1977b); Thoburn (1981a and 1982b) also discusses tin's development effects in other producing countries.
4. This breaks with the generally chronological structure of the book. The analysis of tin exports and development is presented as a single case study in the belief that it would be repetitious to include an 'exports and development' section in each 'chronological' chapter. Also, information is sparse on this topic for earlier periods, with the partial exception of colonial Malaya. Essentially, most of the effects cumulate up to the 1970s, and are best dealt with in Chapter 5 (on the 1960s and 1970s), along with the policy changes introduced during that time. After the 1970s, the mineral rents generated by tin decline rapidly, and the only new 'development effects' consist of the growth of tin-using industries. The latter are discussed in Chapter 5, and there is further discussion of

tin exports and development in Chapter 7. The case study is a shortened and revised version of Thoburn (1981a, pp. 89–113).

5. See Hosking (1988) for a more detailed discussion of types of tin deposit.

6. The 'real' price of tin in Figure 1.1 is that to consumers. The price is deflated by an index of the wholesale prices facing consumers. Before the first world war, when Britain was the largest consumer of tin, the tin price used in Figure 1.1 is the Straits price of tin denominated in £ sterling deflated by a UK wholesale price index. Thereafter, when the US had become the largest consumer, it is the New York tin price (in dollars) deflated by a US wholesale price index. For a discussion of the relationship between the real price facing consumers and that facing producers see the Appendix, and also note 8 below. For tin prices before 1860, see Schmitz (1979, p. 296).

7. See Ayub and Hashimoto (1985, p. 69), who cite indices for the prices (relative to tin) of aluminium, copper, lead and zinc, from 1800 to 1982. These show a consistently rising *relative* price of tin. Over this long period tin has always had a higher price per tonne than lead and zinc. It was cheaper than aluminium, and no more expensive than copper, at least until 1900. As Figure 1.1 shows, the real, long-term trend price of tin was roughly constant over the twentieth century until the 1960s. Tin's major rival, aluminium, experienced a falling real price trend over the same period (Humphreys, 1982, p. 217).

8. Note, though, that UNCTAD uses as a deflator the UN index of export unit values of manufactured goods exported by developed market economies. This means that the real price UNCTAD calculates is a measure of the net barter terms of trade facing tin exporters, rather than the real price to consumers as shown in Figure 1.1. (See also note 6 above, and the Appendix on the relation between the tin terms of trade and the real price of tin to consumers.)

9. After 1985 some statistics on mining methods were still published by the ITC, but were less comprehensive.

10. *TI*, September 1986, gives a fuller account of the technology of bucket wheel dredges, from which the present account is taken.

11. See DMR (1990) for Thailand, and Mines Department, Malaysia, *Monthly Statistics on Mining Industry in Malaysia* (undated, c.1992).

12. Throughput was measured in the past in cubic yards, and still is often referred to as a 'yardage'.

13. This compares to about 0.25 kg in Thailand, and 0.20 kg in Indonesia. Thai and Indonesian offshore dredges worked with similar grades to those of Malaysian dredges, while offshore operations worked with 0.36 kg in Indonesia and 0.23 kg in Thailand. Note though, that the differences are understated (as also in gravel pumping) by the fact that Malaysian operations refer only to ground actually mined, whereas Thai and Indonesian operations include barren overburden stripped before the ore is reached as well as the tin-bearing ground itself (Engel and Allen, 1979, p. 84).

14. Grades in Indonesia in 1989 were estimated for each of P. T. Timah's producing units as: 0.323 kg metal per cubic metre for

Bangka, 0.185 for Belitung, and 0.276 for Singkep (*Mining Journal*, 1991, p. 102).

15. See also Lanning and Mueller (1979, ch. 15), who give a useful description of the increased scale of mining operations in other metals.

16. These small dredges give throughputs of under 80,000 cubic metres a month, very small compared to the large capacity dredges of South-East Asia. Very small dredges, when tried in Asia, have encountered serious problems, being unable to deal for example with heavy vegetation or wood in the material being mined (Thoburn, 1981a, pp. 45 and 57). B. C. Engel (1980), when visiting Brazil in 1979 on behalf of the International Tin Council, reported the first use of a bucket wheel dredge of 100,000 cubic metres per month capacity, then under trial, which had been imported from the United States where it had been developed for harbour excavation.

17. On my field trip to Brazil in 1992 I was unable to collect data on the investment costs of any tin projects.

18. For example in 1980, the year prior to the Malaysian manipulation of the international tin market, only two individual tin companies in Malaysia (Berjuntai Tin Dredging and Malayan Tin Dredging) produced an annual output in excess of 3000 tons of tin-in-concentrates. Only one other individual company in Malaysia produced over 2000 tons, though Koba Tin, a private company in Indonesia, in the same year produced over 5000 tons (*TI*, September 1981). Since the tin collapse, the output of most Malaysian mines has contracted and Berjuntai, the largest tin operating company of the Malaysia Mining Corporation, was producing in 1990 under 3000 tonnes (KLSE, 1990). By the late 1980s, Paranapanema, the largest Brazilian tin company, was producing in excess of 19,000 tonnes of tin per year, mainly from the Patinga deposit (Paranapanema, 1989).

19. In addition to these rents, in the case of minerals there is also a scarcity rent in the sense of an additional surplus of price over marginal extraction cost. This rent represents a kind of 'depreciation allowance' for the mineral deposit which is being depleted (Thoburn, 1981a, pp. 20–3). It used to be argued that scarcity rent would increase steadily over time at the social rate of discount as the resource was gradually depleted (the Hotelling (1931) rule), but Farzin (1992) has shown that the time path of scarcity rent can be quite complicated, and certainly is not necessarily monotonically increasing.

20. One of the most influential works on taxation is Garnaut and Clunies Ross (1983), which argues that the supply of mining investment to Third World countries is sufficiently competitive that LDC governments can impose 'resource rent taxation' so as to secure the bulk of the rents. This view contrasts with what Garnaut and Clunies Ross describe as the 'North American school' approach (e.g. Smith and Wells, 1975), which stresses the role of bargaining in each particular situation, and changes in bargaining positions over time. For other good discussions of mining taxation, see Walrond and Kumar (1986) and Kumar (1991). On the role of

bargaining in commodity production more generally, see Labys (1980) and Maizels (1984).

21. For surveys of this literature, see Thoburn, 1977b, ch. 3, and Thoburn, 1981a, ch. 2.

22. For further discussion of the RV measure, see Gillis and Beals (1980, ch. 1) and Brodsky and Sampson (1980).

23. Two major qualifications need to be made to the idea that backward and forward linkages represent favourable development effects. In the case of forward linkages, since the output of the supplying industry is by definition here a traded good (tin exports!), the forward linked industry could have developed on the basis of imported tin (as was the case with Brazilian smelting and tinplate). Analagously, with backward linkage, the input supplied needs to be non-tradeable (or difficult to trade, like some engineering products which require close customer contact); otherwise the backward linked industry could have established itself independently as an export industry. Second, and especially where the linked industry developed as a result of government intervention (as in the Brazilian tin-using industries), it is necessary to check that it is socially profitable, in the sense of producing net benefits to the nation when its inputs and outputs are valued at world prices.

24. The entries in this section for Malaysia, Indonesia, Thailand and Brazil are substantially updated versions of the producer profiles given in Thoburn (1981a, ch. 3), which lists some additional sources on the years up to the 1970s.

CHAPTER 2

1860–1914
The Early Development of the Industry

The mid-nineteenth century saw a rapid increase in modern industrial demand for tin – tinplate, solder, ball-bearings, machine parts and various alloys – while older uses such as pewter became less important. It also saw the early phase of increased competition between the old-established British tin industry of Cornwall and the newer producers of South-East Asia. By the end of the century South-East Asia had successfully challenged Cornish dominance, and Bolivia too had started to emerge as a major producer. This chapter starts by looking at the decline in the Cornish tin industry's control over the world tin market. This provides a background to the book's starting point of 1860. The chapter then traces the growth of the industry in South-East Asia; particular attention is paid to the role of Western companies and their competition with local miners, who, in South-East Asia, were usually immigrant Chinese. The chapter concludes with an introduction to the early days of the industry in Bolivia. Although some other countries were also exporting tin before the first world war – Australia, South Africa and Nigeria – they are discussed only briefly in this chapter as part of the wider scene.[1]

THE DECLINE OF CORNISH CONTROL OF THE WORLD TIN MARKET

Reference back to Table 1.1 shows that at the beginning of the nineteenth century world output of tin was very small, less than a tenth of the output of 1913. Most of this small supply came from Britain, the Malay Peninsula and China. The Indonesian Archipelago was also established as a tin producer, although its apparent output was small.[2] In the 1860s the British tin mining industry, centred on Cornwall, was still the world's biggest producer, but there had been large increases from South-East Asia, and Straits tin from Malaya had become important in the European market (Wong,

TABLE 2.1   Production of tin-in-concentrates, by country, 1860–1914

(000 tons)	1860	1870	1880	1890	1900	1910	1914
Austria, Czechoslovakia and Germany	0.22	0.17	0.13	0.11	0.08	0.11	0.12
Portugal and Spain	1.3	0.02	0.001	0.03	0.001	0.04	0.22
UK	6.7	10.2	8.9	9.6	3.9	4.8	5.1
Nigeria	–	–	–	–	–	0.57	4.3
South Africa	–	–	–	–	–	2.1	2.1
Other African	–	–	–	–	–	0.30	0.35
USA	–	–	–	–	–	0.03	0.09
Bolivia	–	0.10	0.36	1.7	9.1	22.8	22.0
India	–	–	–	0.04	0.07	0.14	0.23
China	0.50	0.50	4.0	3.0	2.9	6.4	7.0
Indo–China	–	–	–	–	–	0.07	0.07
Japan	–	0.01	–	0.05	0.01	0.02	0.29
Malaya	7.0	9.0	11.7	27.2	43.1	45.9	50.6
Siam (Thailand)	–	–	3.0	4.4	3.9	4.2	6.6
Dutch East Indies (Indonesia)	5.3	7.3	9.2	12.4	17.6	21.4	19.5
Australia	0.13	0.19	10.4	7.4	4.3	6.8	5.4
World	19.8	27.5	47.7	65.9	85.0	115.7	124.1

Sources and Notes
1. From ITRDC (1937)
2. Countries not producing over 100 tons in any year 1860–1914 have been excluded.
3. Malayan statistics for 1860 and 1870 probably include production from Siam.

1965, p. 17). There was also production in Australia and Bolivia. Australia, briefly, was the world largest producer for some years in the 1870s and 1880s, Cornish output having peaked in 1871 (ITRDC, 1937). By 1913 world output had risen by more than five-fold compared to the mid-nineteenth century, and the geographical pattern of production which was to characterise the industry until the 1980s was established.

The Cornish tin industry dates back to well before Roman times, and for many centuries before the industrial revolution tin was one of Britain's main exports (Hedges, 1964, ch. 1). In the first half of the nineteenth century the global tin industry was segmented, with Cornwall dominating the Western world. The producers in South-East Asia served the Chinese market, whose own tin mining industry could not meet the local demand for tin for uses such as coinage and the manufacture of joss-paper (for religious offerings) (Heidhues, 1992, pp. 3, 23). The segmentation resulted in part from the poor quality of Asian tin, which was mostly still smelted by simple methods, and which produced a metal with too high a level of impurities to be suitable for the new industrial uses. Import protection in the UK was also used to support Cornish tin. Some tin from the Netherlands East Indies (Indonesia) was shipped to Europe, but Cornish tin exporters nevertheless were able to maintain high prices in Europe, where they were the only significant producers, and to sell off surplus supplies to China. The Cornish tin mining industry was highly integrated with British smelting, which reduced competition within the British industry and provided barriers to entry (Hillman, 1984).

Expansion of tin exports from Bangka in the Dutch East Indies had penetrated the European market with metal of an acceptable quality by the mid-nineteenth century, and pressure from buyers resulted in the opening of the British market. Tin exports from the Straits Settlements shifted towards Europe at about this time with the growth of industrial consumption in Europe, especially from the tinplate industry, following the successful mechanisation of the hot-dipping of tinplate in 1856 (Robertson, 1982, p. 8). American consumption also expanded, and by the end of the nineteenth century the US was generating about a quarter of world demand (Smith, 1980–2). Despite the breaking down of market segmentation, Cornwall's underground mining remained able to compete until the 1870s, with increased investment financed by smelting interests. In 1871 the Cornish mines produced their largest output (11,320 tons) since production figures were first recorded in 1670 (Smith, 1980–2). From the 1870s, as Malayan production started to expand (see Table 2.1) from low-cost alluvial deposits, Cornwall declined in relative importance. Australian production, started on a significant scale in 1871, also was an important new source of supply, but its production declined after the early 1880s. The depression of the late 1870s damaged the Cornish industry. When tin prices fell in the 1890s (see Figure 1.1), the Cornish mining industry went into a decline from which it never recovered. By the first world war Cornish tin output was confined to a few large mines, a situation which persisted into the 1980s (when all but one mine was closed). Until the development of smelting capacity in Malaya towards the end of the nineteenth century, most of the world's tin was smelted in Britain. The UK remained the world's largest producer of tinplate until 1912, although after the imposition of import protection by the US in 1891, American tinplate production grew rapidly (Minchinton, 1957, ch. 2).

The decline of Cornish mining was accompanied by an *expansion* in British smelting, as new smelting firms were established. Overcapacity in smelting was solved in part by amalgamation. In the face of intensified competition, smelters looked for overseas sources of ore (Hillman, 1984, pp. 411–20). As seen later in the chapter, this demand for ore influenced the growth of the Bolivian tin industry from the 1890s. Bolivian ore, which comes from hard rock, is much more difficult to smelt than the alluvial tin concentrates of South-East Asia, making a local smelting industry in Bolivia harder to establish. Cornish capital was also important in the early development of Western involvement in the tin industry in Malaya. In general there was a strong interest in Britain in overseas mining investment – for example there was a boom in Nigerian tin shares in the years shortly before the first world war (Wong, 1965, p. 215). Tin was also discovered in the Transvaal in South Africa in 1904 (Jones, 1925, p. 290).

THE GROWTH OF THE TIN MINING INDUSTRY IN SOUTH-EAST ASIA

In none of the three main producers in South-East Asia did tin mining depend for its early growth on foreign firms. In Malaysia and Thailand Western firms initially faced strong competition from well-established mines run by immigrant Chinese. In Thailand, the struggle between Chinese and Western companies had a strong political dimension, and the development brought about by immigrant Chinese miners operating in the south of the country was an important bulwark against prospective British colonial incursion. In Indonesia much of the industry was developed by the State; again, however, there was a prior history of Chinese involvement in tin mining, and the mining during the nineteenth century and early twentieth centuries was done by Chinese under indirect control.

Malaysia and Thailand[3]

At the time of British intervention in Malaya, the country was almost entirely covered in jungle. Settlement was confined to Malay communities along the coast and on the banks of rivers, the Chinese mining community of Larut in Perak, and a number of Chinese agriculturalists practising shifting cultivation of such crops as pepper, gambier and tapioca (Jackson, 1968, pp. 34–83). Already long-established in the Straits Settlements of Singapore, Penang and Malacca, British control was extended to the Malay Peninsula proper in 1874. In that year the Sultan of the west coast state of Perak accepted under the Pangkor Treaty a British adviser, or 'Resident', whose advice he was obliged to accept on all matters except Malay custom and religion. This form of indirect rule was soon extended to other Malay states, and in 1895 the Federated Malay States were formed, comprising Perak, Selangor, Negri Sembilan and Pahang. Soon after, all states in the Malay Peninsula were under British control, Perlis, Kedah, Kelantan and Trengganu having been ceded from Siam. The whole area was known as British Malaya, comprising the States Settlements, the Federated Malay States (FMS) and the Unfederated Malay States (UMS). Only the Straits Settlements were under direct rule.

The early development of the economy, in the second half of the nineteenth century, was closely connected with tin (Wong, 1965; Yip, 1969). Tin had been produced in Malaya since the ninth century (Fermor, 1939, pp. 21–3), but its large-scale exploitation dates from the discovery of the Larut tin field in 1848 and the consequent influx of Chinese miners. By 1870, 40,000 Chinese were mining tin in the district (compared to a peak mining force in the Federated Malay States as a whole of 225,000 in 1913), and indigenous Malay miners had long since been pushed out of business. Disputes between rival Chinese clans, culminating in open warfare, led to an exodus of population and so a large fall in tin output; they also provided a pretext for the intervention of the British. By the end of 1874 output had recovered and many of the Chinese who had left the district returned. Exports of tin from Larut rose rapidly, reaching a peak of approximately three-quarters of Perak's output in 1884. In Selangor too tin was being mined, with 7000 Chinese working in the mines around Kuala Lumpur in 1874. This rose to 23,000 by 1884. Three-quarters of Selangor's output came from the Kuala Lumpur area. Malaya soon overtook Britain, the then Dutch East Indies and Australia to become the largest world producer by 1879. By 1896 Malaya produced about 60 per cent of world output.

Tin was discovered in the Kinta Valley around the city of Ipoh in Perak in the early 1880s (Loh, 1988, p. 7). It began to rival Larut as the main tin area by the mid-1880s, and was to remain the centre of the tin industry for most of the twentieth century. In 1885 the Resident of Perak noted an extensive migration of Chinese miners from Larut to other parts of the state, which would have been mainly to Kinta. By 1889 Kinta's output surpassed that of Larut, and in 1910 its output was nearly seven times that of Larut, and approximately 40 per cent of the output of the Federated Malay States. Indeed, this was the start of a strong regional pattern in tin mining which continues to the present day, with the bulk of output coming from Perak and Selangor.

Until the last years of the nineteenth century tin mining was almost exclusively a Chinese concern, in part because of local Chinese control over merchant capital. Capital for the industry came initially from the Chinese merchant communities of the Straits Settlements, and apparently very little was from China (Wong, 1965, pp. 60–4). A system of advances allowed the cost of fixed and working capital to be shared between a mine owner and an advancer. Use too was made of trade credit, with a pyramiding of facilities from large to small suppliers to miners, a system still in force in the 1960s (Tan, 1966). Fixed capital costs were low because little equipment was needed for opencast mining. Working capital costs were kept down by the practice of not paying workers until the end of six months or even a year; many workers worked on 'tribute', a system whereby they 'lent' their labour to the mine owner in return for a share in the eventual profits of the mine. Additional profits were made by advancing food and opium to mine workers and by the leasing of revenue 'farms' from the government, which gave the holder the right to levy taxes on a particular commodity or activity, such as opium or gambling, in return for a fixed cash payment. Until the coming of rubber, tin mining was the major investment opportunity in the country and it seems likely that the rapid expansion of the industry from 1874 to the depression years of the 1890s, was in large part financed through reinvested profits, including those of advancers.

The labour force of the Chinese mines was drawn directly from China. Workers first came mainly under an indenture system, whereby their passage was paid by prospective employers, with whom they had to work off their debt. Later, workers were drawn from a pool of people who had paid off their indentures or who came from China under their own finance. These workers formed the core of the Chinese population of Malaya, who at the time of Independence would constitute about a third of the country's total population. By 1914, when the indenture system was officially prohibited, it had virtually died out (Li, 1955, p. 86). Participation by Malays as either owners or workers was practically non-existent. Malays apparently were unwilling to work in the exacting conditions of the mines, at least for the prevailing rates of pay. In any case differences in language and social custom, the activities of Chinese secret societies and the feeling of local Chinese public opinion made their employment difficult, except as clearers of the jungle prior to the establishment of a mine (Wong, 1965, pp. 64-5).

At the turn of the century, when Malaya was already supplying over half of world tin output, Western firms were only just beginning to make their first inroads into Chinese predominance in tin (Wong, 1965, p. 53). Apart from the Pahang Corporation, an underground mine which was the sole survivor of a 'tin rush' in Pahang in the late 1880s (Wong, 1965, pp. 137-41), most Western firms had been unsuccessful in their attempts to establish themselves in Malaya. While easily accessible high-grade ground was being mined by labour-intensive Chinese methods, Western firms proved unable to compete, being encumbered by a top-heavy administrative structure and inappropriate equipment (Allen and Donnithorne, 1954, pp. 150-2). There were many company flotations, but there was also a high failure rate. Moreover secret society pressures made it difficult for them to hire Chinese labour. By the end of the 1890s, however, Chinese firms were experiencing difficulties. Chinese mines were open pits and used very large labour forces, roughly ten times more workers being employed per unit of throughput than in a gravel pump mine of the 1970s (Thoburn, 1977b, p. 127). Although steam pumps had been introduced in 1877, there were still drainage problems and the difficulties of raising ore-bearing earth to the surface from deep pits made greater use of mechanical equipment more appropriate (Wong, 1965, pp. 216–21). The mining of lower grades of ground following the exhaustion of the richest deposits gave an edge to methods with a higher throughput of tin-bearing ground, and there was an increasingly severe labour shortage. This was partly the result of the tin market recession from 1889 to the late 1890s, which reduced immigration, and also factors temporarily reducing emigration from China such as a good harvest in 1897 (Wong, 1965, p. 172). Thus, the search for labour-saving techniques became important for the whole industry. It was intensified from 1896 to 1907 by the authorities' closure of the land application books in the Kinta Valley (Wong, 1965, pp. 114, 220), which diminished opportunities of working new land with traditional methods. Also the new rubber industry competed to some extent for labour from the first decade of the twentieth century, though much of its demand was met by importing Indian workers. The start of the abolition of the practice of revenue 'farming' from 1909, which was completed by 1912, deprived Chinese mine owners (who were usually involved in other forms of making money too) of lucrative sources of revenue, as did also the decline in the demand of their mine workers for opium (Wong, 1965, p. 222).The suppression of Chinese secret societies by the government by this time also was an important factor in promoting the growth of European firms in relation to Chinese ones.

Western firms were now at an advantage since they had better access to new techniques used to mine other minerals elsewhere. Thus gravel pumping was used in Cornwall, the centre of tin mining in England, for mining china clay (Allen and Donnithorne, 1954, p. 157) and precious metals were mined in Australia and New Zealand by dredging. What is interesting is that the first major new technique – gravel pumping – which helped to establish prominent British companies such as Osborne and Chappel (the operators of the Gopeng mine) in the industry, should have been taken up by the local miners and has formed the basis of the local sector in Malaysia since about the 1920s to the present.

The first hydraulic mine was the Gopeng mine in Perak in 1892, a British company with strong Cornish connections. This was the precursor of gravel pumping, but it used a natural head of water rather than mechanical power. The first use of artificial motive power to operate the jets and elevators came in Malaya in 1906. This was the so-called suction dredge. Pumps mounted on a floating pontoon, operated a monitor to break down ground on the wall of the excavation, and a centrifugal gravel pump elevated the slurry. Because of its high operating costs the method never became popular but its importance lies in the fact that it demonstrated the possibility of hydraulic mining operating with artificial power (i.e. gravel pumping), and it was rapidly adopted by Chinese miners who adapted it for use without the floating pontoon. By 1915 the Malayan Mines Department's details of principal workings in the Federated Malay States listed three mines, all in Perak, hydraulicing with power plant; by the 1920s gravel pump mining was well established as the main form of Malayan Chinese mining.

Bucket dredges, in contrast, required large capital outlays which the Chinese system of credit could not finance. The European dredging firms were joint-stock companies with capital raised from abroad. Capital from Cornwall was an important source of finance, and the Osborne and Chappel group was closely linked with the Redruth Mining Exchange in Cornwall (Wong, 1965, p. 216). Since then, once dredging had become established in Malaya, dredge design was available from specialist consulting firms, and as dredges could be built by independent shipyards and engineers, it seems that capital rather than access to technology has been the crucial constraint on Malayan Chinese entry into dredging (Hennart, 1986a, pp. 136–7). Chinese businesses may also have been inhibited from the long-term investment required in dredging by fear of official discrimination against them (van Helten and Jones, 1989, p. 166). Interestingly though, as will be shown later, influential local Chinese played a significant role in the introduction of the very first dredge – in southern Thailand in 1907.

Although government policy in the years immediately before the first world war was said to favour Western firms (Wong, 1965, pp. 220–1; van Helten and Jones, 1989, p. 167), the British-controlled governments of the individual Malay states had taken various measures to encourage mining during the period of Chinese dominance, the export duty on tin being the main source of government revenue. Most of the major towns on the west coast grew up as mining centres – Ipoh in the Kinta valley, Taiping in Larut, Kuala Lumpur in the biggest Selangor tinfield and Seremban in Negri Sembilan – and the government used tin revenue to construct railways from these centres to seaports, whence the ore could be shipped to Penang or Singapore for smelting. In 1885 eight miles of track from Port Weld to Taiping were built by the Ceylon government under contract to the British authorities. Kuala Lumpur was connected to the port of Klang in 1886, Seremban to Port Dickson in 1891 and Ipoh to Telok Anson in 1895. By 1910 a line had been built from Prai (on the mainland by Penang Island) to Johore Bahru, and extended over the causeway to Singapore in 1923. This connected all the main mining centres and constitutes the basic west coast railway system of the present day. Such railways not only facilitated the growth of tin mining, but opened up the country to further settlement and development. The importance of communications was clearly seen by the government, which was also keen to develop the mining industry as a source of tax revenue. Thus, F. A. Swettenham, as Resident of Selangor, declared in his 1883 annual report that:

> The government of Selangor has been anxious to improve, or rather to make, communications between the various states, to bring the mines within reach of the bases of supply, and to induce immigrants to come into the country and plant, giving them good roads to a market.

Finance for railway construction came mainly from current government revenue, and much of this in turn came from export duties. During the peak of railway construction from 1884 to 1910, tin export duty generated 36 per cent of total FMS revenue (Thoburn, 1977b, pp. 123 and 130).

Until the development of rubber, from 1905, tin mining was the principal source of growth in the Malayan economy. Although the economy of the indigenous Malays benefitted from a ready market among the Chinese miners for the Malays' fruit and livestock, the output of the main Malay crop, rice, did not expand to meet the potential demand, in spite of help by the government for irrigation schemes (Chai, 1964, p. 147). Chai attributes this failure to the lack of widespread irrigation and the primitiveness of the existing agricultural equipment. Malay participation in the export sector had to await the coming of rubber.

In Thailand ('Siam' before 1939) the political economy of tin production during crucial phases of its development revolved not only around the competition between ethnic Chinese and Western enterprise, but that between Britain as a colonial power and Siam as an independent country. In the mid-and late nineteenth and early twentieth centuries the Siamese state had relied heavily on prominent Chinese to administer the southernmost provinces of Thailand and to supply tax revenue from those provinces. This provided a degree of indirect control by the state over remote areas with poor communications to the centre of power. By developing the southern provinces, mainly through tin mining and related activities, the Siamese government was to some extent able to deprive the British of a pretext for intervention – a risk illustrated by the comments of an official in 1906 that: 'The British Government will insist that the country shall be opened up either by Siam, or, if she be unwilling, then by Great Britain herself.' (Cushman, 1991, p. 47).

In southern Siam immigration by Chinese in response to mining opportunities had been taking place for several hundred years prior to the industry's expansion at the turn of the twentieth century, and Thais themselves rarely mined. Chinese from Malaya were actually invited to invest in tin mining by southern provincial governors in the late nineteenth century in order to raise tax revenue (Wit, 1971, p. 9). In the 1890s the tin districts were brought within a new administrative structure, with a high commissioner appointed from Bangkok, and in 1891 a Department of Mines was established under *British* officials (Falkus, 1989, p. 148)

In the last quarter of the nineteenth century, after an earlier period of expansion from 1850 to 1870, the tin industry in Siam stagnated according to Falkus (1989, p. 147). He argues that increased production from Malaya pushed down prices, as well as competing for ethnic Chinese labour and capital. In fact, as Figure 1.1 shows, after an initial fall from the early to the late 1870s, real tin prices recovered and rose until the late 1880s; it was only from the late 1880s to the late 1890s that prices fell.[4] As costs began to rise in Malaya in the early years of the twentieth century, and tin prices began to recover sharply from the late 1890s, Western companies developed interest in Siam as a source of tin (Falkus, 1989, p. 147). Siam, however, remained reluctant to admit Western companies and was deeply suspicious of the British.

Western involvement in the Siamese tin industry before the first world war faced many difficulties, and started later than in Malaya. Nevertheless, by 1913 western (in this case Australian and British) companies were producing 34 per cent of Siamese tin output (Falkus, 1989, p. 153), a higher figure than in Malaya where the share was 26 per cent (Yip, 1969, p. 149). As in Malaya, the success of Western firms depended on the development of dredging. Two main groups were involved in Siam. One group was started by the Australian dredging pioneer, Captain E. T. Miles, in collaboration with politically powerful local Chinese mining interests. Miles appears to have been acceptable to the Siamese authorities in part because he was not linked to British business and political interests; thus the Siamese could 'show willing' with regard to allowing Western business involvement while minimising its political implications (Falkus, 1989, pp. 153–4). The local Chinese interests who supported Miles were the members of the *Khaw* family (*Xu* in modern Chinese (Pinyin) romanisation), whose influence on the development of the Siamese and Malayan tin industries has been chronicled by the research of the late Jennifer Cushman (1991), which is discussed later in this subsection. Miles, who had earlier business contacts with the Khaws when he supplied several vessels to their shipping company, was invited by a prominent Khaw who was then High Commissioner of the Phuket area to develop a new tin property at Phuket 'by modern European methods' (Cushman, 1991, pp. 69–75). Out of this invitation grew the Tongkah Harbour Tin Dredging Company, still active in the early 1990s, which started work with a bucket dredge purchased by Miles from Scotland. This was the first dredge used to mine tin in Asia.[5] Miles had seen dredges mining gold in New Zealand, and with various adaptations made by him during the early days of Tongkah Harbour's operations, the technology was successfully adapted to the mining of tin, in this case with the dredge working offshore. Although Miles and other Australian interests had substantial shareholdings, there were also 'Eastern shareholders', who were represented on the board of directors by a member of the Khaw family. Other companies were started by Miles and worked in the Ranong area some 120 miles north of Phuket.

Whilst Miles and his associates appear to have been unconnected with Western interests from Malaya, tin mining in Siam was also started at an early stage by companies floated by Guthries, the well-known agency house from British Malaya, though even Guthries' interests in tin in colonial Malaya were not of great importance. These companies – the Renong Dredging Company and the Siamese Tin Syndicate (the latter of which was still operating tin in the 1980s) – operated dredges in the Ranong area. The Renong Dredging Company's dredge, which started work in 1910, was the first land dredge to mine tin in South-East Asia (Falkus, 1989, p. 152). The date of introduction of gravel pumping in Siam is not known (Pajon Sinlapajan, 1969). It does not seem to be known if gravel pumping was introduced to Thailand by foreign companies either. More likely it was introduced by Chinese from Malaya. In any case, the degree of competition between foreign and local mining firms before the first world war would have been less intense in Thailand than in Malaya since many of the first foreign dredging companies were operating offshore and therefore not competing for leases.

The role of ethnic Chinese in the development of tin mining in Siam, their influence in activities in related areas, and their competition with Westerners, is well illustrated by Cushman's history of the Khaws. Leading members of the Khaw family operated revenue 'farms' for opium and other activities, under which they delivered tax revenue to the state in return for the right to levy taxes on the activity; they acted as governors or other leading officials in the tin producing areas, particularly Ranong and Phuket.

The Khaws, who maintained strong social and economic links with other leading Chinese business families in both southern Siam and in Penang in Malaya, had good connections with the Siamese court in Bangkok. They had first moved into the tin business through shipping, not only of tin ore but of immigrant Chinese labourers from Malaya, and opium and other supplies. Local political power gave them control some control over the issue of mining leases, particularly in Phuket, Siam's main tin mining area.

In 1887 the Straits Trading Company (STC) had been formed by Western interests to smelt tin ore from Malaya. The STC offered miners the attraction of cash advances for their ore, and set up modern smelters off Singapore and Penang. The company had close associations with the Straits Shipping Company, which carried ore and mining labour and supplies. By the beginning of the twentieth century, the Straits Trading and Straits Steamship companies wanted to expand into the tin markets of southern Thailand. Cushman (1991, p. 52) records that the Khaw family, whose interests were directly threatened by this proposed expansion, were encouraged by the Siamese government to develop their enterprises in order to counter the incursions of British interests.

That the British colonial government in Malaya was willing to help British smelting interests is illustrated by the imposition of a prohibitive duty on the export of tin ore from 1903, from which exports to the UK were exempted (from 1904). This was designed to prevent an American attempt to establish a smelter in the United States using Malayan concentrates.[6] This attempt was by one of the largest buyers of tin cans in the world, Standard Oil, which used them to ship kerosene. Faced by a highly monopolised domestic supply of tin cans from the US steel industry, protected from import competition by a tariff, Standard Oil had wished to increase its bargaining power by integrating backward into smelting. A similarly prohibitive duty was imposed in 1915 on sales of Nigerian tin concentrates to non (British) Empire smelters to block sales to the American smelting industry developed during the first world war (Hennart, 1986b, pp. 228–9, 266–7).

The Khaws were able to join with other ethnic Chinese interests to form a vertically integrated structure to compete with the British. The family group's financial strength derived from years of profits from revenue farming, as well as profits from its shipping and mining activities. Tin not smelted by the Straits Trading Company tended to be smelted by small Chinese-owned smelters in Thailand, which had low capital costs but produced highly impure tin metal. A small, modern smelter was set up in Penang by an associate of the Khaw group, and was purchased in 1907 by a new company, the Eastern Smelting Company, owned mainly by ethnic Chinese mining interests among whom the Khaw group was prominent. However, it was open to public subscription and there were influential Europeans on the board of directors. Eastern Smelting soon opened a network of ore-buying agencies throughput the Federated Malay States and was competing effectively with Straits Trading. By 1910, Eastern Smelting was smelting some 29 per cent of total shipments of tin from the Straits Settlements (Cushman, 1991, p. 79). Eastern Smelting's competitive power was complemented by the earlier strengthening of Khaw shipping interests to form the Eastern Shipping Company, which had been floated in 1907 in Penang, with a board of directors representing a wide cross-section of the Peninsula's ethnic Chinese mining and business interests. Both the Straits Trading Company and Eastern Smelting (later renamed Datuk Keramat) exist today and operate smelters.

Cushman records a gradual decline in the economic power and influence of the Khaw family just before, and in the years after, the first world war. The Siamese state's fear of potential British extension of power into the southern provinces lessened after a treaty signed in 1909, and the need to rely on the local Chinese community to maintain a buffer against Malaya was lessened. Under the 1909 treaty Siam ceded its former tributary states of Kelantan, Trengganu, Kedah and Perlis to British Malaya, and guaranteed it would not allow any other power 'to establish itself in that portion of the Malay Peninsula which remained under Siamese control' (Cushman, 1991, p. 95). After a prominent Khaw was murdered in 1913, he was replaced as High Commissioner in Phuket by a Thai. After 1910, it was said that applications by Western companies were viewed more favourably. Although the Khaws were influential in helping some Western mining companies (those associated with Miles) to establish in Siam, their power in relation to Western interests was reduced as the Western dredging sector gathered momentum, with the backing of home governments and financial institutions (Cushman, 1991, p. 114). The Eastern Smelting Company suffered from a lack of capital for expansion and was sold to Western interests in 1911. Eastern Shipping, whose ships were requisitioned by the British during the first world war apparently without adequate compensation, was sold to the Straits Shipping Company in 1922.

Indonesia[7]

Dutch colonial interest in South-East Asian tin can be traced back to the early period of European involvement in the region. The Dutch East India Company inherited a well-established tin trade when it took over Malacca in 1641; when tin was discovered on Bangka island at the beginning of the eighteenth century, the Company was keen to develop the trade. In South-East Asia in the late eighteenth century Bangka island became one of two major centres of tin mining, the other being Phuket (known at the time as 'Junk Ceylon') in Thailand. In the 1760s output was thought to have been about 3500 tons, roughly the same as that of Cornwall at the time, and mostly exported to China. There was particularly a strong demand for tin by China, beyond the country's own mine production. The potential of the tin trade to China also lay behind Britain's interest in Bangka island during the brief period of British rule of the Indonesian Archipelago from 1812–16, after which the area was ceded back to the Dutch.

By 1860, the Dutch colonial administration of Bangka, which also administered the tin mines until the separation of this function in 1913 to a separate state agency, was operating a system of indirect rule over the industry, essentially one inherited from the brief British period. Mines were organised as cooperatives (*kongsi*),[8] with workers recruited almost entirely from China. There had indeed been a large Chinese presence on the island for more than a century, Chinese miners, as in Malaya, having long since pushed local miners out of the industry. Initially the kongsis had operated to some extent as genuine cooperatives, with members working in the mine, but increasingly as the century progressed, shareholders ceased to be workers, and workers were recruited under a form of indenture. Heidhues' work chronicles the poor conditions of these workers, who were often in long-term debt, and prevented from leaving their employment by harsh penalties imposed by the Chinese mine managers and the colonial administration. Workers, it seems, were significantly worse off than in the Malay Peninsula, which was able to attract many migrants who might otherwise have gone to the Dutch East Indies. Few migrants were able to return to China with savings. The government of China expressed worries from time to time about its nationals' working conditions.

Chinese mine operators acted as a buffer between the Dutch and the mine workers, and linguistic barriers tended to prevent direct Dutch involvement with the workforce. Mines were paid a fixed price for tin, which did not rise, at least in the short term, with the export price of tin, although mine operators still could make profits by their control over sales of food and opium to mineworkers. The Dutch advanced capital for mining supplies and wages, and over the nineteenth century the Dutch secured increasingly tight control over mining operations. To some extent, the present day use of subcontractors by the Indonesian state tin company, P. T. Timah, can be seen to have its origins in the nineteenth century.

Tin mining on Belitung island was administered from 1860 by a Dutch company, the Billiton Company, whose successor is still in existence today as a subsidiary of Royal Dutch/Shell. Tin had been discovered about a decade previously, and a group of western pioneers in 1852 had been granted a forty year concession, which was soon taken over by the Billiton Company. Like Bangka, the company worked its mines indirectly, with the mining carried out by Chinese. The third of present-day Indonesia's three tin mining islands, Singkep, started production in 1891.

The Billiton company started as a privately-owned concern, but after the company ran into financial problems in the early 1890s with the low tin price, the state acquired a five-eighths share in 1892 when the first lease was renewed (Heidhues, 1992, pp. 80–1).[9] Another Dutch company worked Singkep (Jones, 1925, p. 233), but its concession would be taken over by Billiton in 1933.

Most of Bangka's ores were smelted on the island until the twentieth century, while some ore from Belitung was shipped to Singapore.

The ability of the Dutch to use Chinese labour for the highly labour-intensive open-cast methods of the nineteenth century contrasts with the experience of the British mining companies in colonial Malaya. Indeed, in this regard the Dutch were continuing a tradition of employing Chinese labour that was used by the local rulers before them (ter Braake, 1944, p. 36). Possibly the greater power of the Dutch in relation to Chinese miners sprang from their control over tin marketing, and the greater control it was possible to exercise on small island societies.

On Belitung island the company predominantly had workers working under subcontractors using a 'tribute' system, and workers sold ore to the company with certain guarantees about minimum payment. Recruitment was organised for a while by the company itself but by the end of the nineteenth century it was left to relatives of existing workers.[10] Unlike Bangka, where the early mining kongsis were formed by men (there were few women) with kinship or regional links, those on Belitung were organised out of workers with no such connections, and there was little emphasis on shareholding (Heidhues, 1991). Nevertheless workers' standards of living on Belitung were higher than those on Bangka, and workers on Belitung were not subject to the penal sanction which was used on Bangka until well into the 1930s, which made it a criminal offence for mineworkers to leave their employment. These higher standards seem to owe much to a system introduced by the company of organising kongsi workers into teams (or *numpang* – an Indonesian term meaning to join with others), whose members shared the profits of the operation, with the shareout depending on how well workers fulfilled worknorms. The shareholders in the kongsi also shared the profits, but in a fixed proportion agreed with the numpang. One kongsi, owned by a small number of shareholders, might operate several such numpangs of workers. Not all workers were in numpangs – new arrivals were indentured until their passage money had been repaid were excluded, as were workers on smaller Chinese mines not administered by the company (Heidhues, 1992, pp. 79, 125–6).

The techniques in the Dutch East Indies were those of large, open-cast mining similar to those of nineteenth century Malaya; the Malayan Chinese organisation of the labour force in kongsis and of financing by advances was also antedated by Bangka. These Chinese mining methods, which required sophisticated labour organisation compared to the small-scale mining which had preceded it, also had introduced new technology in the form of the water-driven chain pumps which allowed mining down to depths of about twenty feet. The widespread use of mechanical power only got underway by the end of the nineteenth century however. Techniques changed as lower grades of ore were encountered after the initially rich deposits were worked out, and there was an increasing labour shortage as the nineteenth century progressed, exacerbated in the case of Bangka by its bad reputation among prospective workers. On Billiton island by the early 1880s fears were being expressed about the exhaustion of the island's reserves and an intensified exploration programme was started, although from 1875 to 1891 Belitung's output had risen to rival and sometimes to surpass Bangka's. On Bangka there had been difficulties in finding both new sites and adequate labour from the mid-nineteenth century on.

The suction dredge was introduced from Australia and the Straits in 1906 and, as in Malaya, the floating pontoon seems quickly to have been dispensed with and the technique was developed as gravel pumping. On Bangka the first bucket dredges were not operated until 1927 (Heidhues, 1992, p. 128). The first bucket dredge on Billiton dated from 1920, and a sea dredge was floated off Singkep in 1911 (Zaalberg, 1970). Dredging would later spread quickly and by 1939 dredges produced 40 per cent of the country's output (Fox, 1974, p. 32).

Thus in colonial Indonesia there was no interplay between choice of technique and ownership and control. The Dutch colonial administration on Bangka, together with the Billiton company and the Singkep company, had been given an effective monopoly of tin mining; so gravel pumps were used primarily to mine small deposits.

THE EARLY DEVELOPMENT OF THE INDUSTRY IN BOLIVIA[11]

Tin mining had been developed in Bolivia by the Incas, but little concern was shown for the metal by the Spanish conquerors, who were more interested in the country's silver deposits. Bolivian mineralisation is such that tin and silver often occur together. Since tin had been mined as a byproduct of silver, the well-established silver mining cities of Oruro and Potosi on the high plateau formed a nucleus for the growth of tin. Tin production in Bolivia grew rapidly in the 1890s, coinciding with the exhaustion of the main silver deposits and a drop in the market for silver, hitherto the largest export. Much of the basis for expansion had been laid from the 1860s, however. British smelters had taken a strong interest in buying Bolivian ore. New entrants who moved into smelting tin in the 1880s, such as the then Bristol-based smelter Capper Pass (with experience in other metals), bought Bolivian ores, and the growth of a smelting industry in continental Europe from the 1890s also increased competition for ore. When the price of silver fell in the 1890s, this was equivalent to a depreciation of the Bolivian currency, and helped to boost tin exports, while damaging the prospects of silver mining. Growth was aided in the 1890s too by Oruro being linked by rail to the Chilean coast. This not only improved ore transportation but subsequently provided a source of mining labour in the form of former railway building workers (Hillman, 1984).

The development of the Bolivian tin industry was aided by the fact that, from the 1880s, the government was closely allied with mining interests.

The tin industry was more lightly taxed than in Malaya or Cornwall. Silver mining had created an extensive infrastructure of finance and commerce, and there was already a foreign presence, with investors from Chile, France, Britain and Germany. Hillman (1984, p. 433) estimates that some 40 per cent of output by the end of the century was coming from sources which were directly connected with silver mining. Much of the rest were from operations by individuals with silver mining expertise; the main silver mines were foreign-owned, and many of the individual developers were foreign.

This period also saw the beginnings of the career of Simon Patino, the Bolivian who acquired control of La Salvadora mine in 1897, subsequently making a major discovery there which was to form the starting point of his rise to become Bolivia's largest producer. Patino would go on to found one of the world tin industry's most powerful multinational companies, based in the USA after 1924.

From 1905 to 1921 Bolivia was the second largest producer of tin, after Malaya, and vied for second place with the Dutch East Indies for most of the 1920s and 1930s.

OVERVIEW: 1860–1914[12]

The years from the latter half of the nineteenth century, and particularly from the 1870s, to the first world war saw the development of demand for a wide range of raw materials to feed the industrialisation of Britain, Western Europe and the United States. Some of these materials, like tin, would came from colonial economies. The focus of tin supply switched rapidly away from the main traditional producing area of Cornwall in Britain to the countries which are now Malaysia, Thailand and Indonesia. Later in the period Bolivia became an important exporter. Australia for a short time also was a major producer.

Indeed, the need for raw materials was an important motivation for colonialisation. The extension of British colonial power from the Straits Settlements of Singapore, Penang and Malacca to the Malay Peninsula proper was associated with a British wish to restore order to the tin mining areas of Malaya. Siam was fearful that the British would seek a pretext to occupy the south of the country, and the Siamese government encouraged the tin mining industry in those areas as a source of revenue and to show that it itself was prepared to ensure development.

The early development of the tin industry in South-East Asia was closely associated with the economic activities of the region's overseas Chinese, who came mainly from China's southern provinces of Guangdong and Fujian. By the mid-nineteenth century they had driven out of business such indigenous miners as there were. The Chinese, with their ready access to capital (at least short-term capital), and by their control over the labour force by secret society organisation, were to prove formidable competitors to Western companies in Malaya and Siam. It was not until the decade before the first world war, with the development of the technique of tin dredging, that Western companies were able seriously to challenge Chinese dominance. In Siam incursions by Western companies were resisted for most of the period and the Chinese were favoured. Western companies in Malaya were helped in part by the colonial state, which, for example, suppressed secret socities. Even so, the development of the technique of gravel pumping by the local Chinese miners allowed them to survive as a thriving smaller scale sector, instead of being wiped out by the foreign competition. This contrasts, for example, with the experience of the local miners in the copper industry of Chile in the early twentieth century, whose business did not survive the competition of American companies. In the Dutch East Indies, although Chinese comprised the bulk of the labour force in the tin-mining islands, and Chinese entrepreneurs acted as mine contractors, the industry was firmly in the hands of the Dutch colonial state. Even the one foreign company was Dutch, and the colonial government shared in its ownership. In Bolivia, although foreign capital from the United States and Chile was present, the industry's most famous company, and the forerunner of a powerful multinational, was that of a Bolivian, Simon Patino.

The Western companies in Malaya and Siam mainly were British, and in some cases Australian. Some British companies had connections with the tin mining industry of Cornwall. American companies did not gain any significant foothold; neither mining companies nor tin-using companies. Britain prevented the development of American smelting by a prohibitive duty on exports from Malaya of unsmelted tin ore, and later on those of Nigeria too. This echoes the way in which American tyre interests were prevented from securing estate acreage in the Malayan rubber industry.

Tin prices during the 1860–1914 period showed great fluctuations, as Figure 1.1 illustrates. While these certainly affected the pace of economic activity in tin mining, the apparently low level of real prices in the nineteenth century, in comparison to the prices of the twentieth, were compensated for by ore grades which were exceptionally high by today's standards.

Of the important South-East Asian producers, only in Malaya was tin dominant in export earnings, but by the early years of the first world war rubber export earnings were larger than those from tin. Siam's export earnings were far more heavily dependent on rice than on tin, and the country also had important teak exports, as well as eventually developing rubber. When the tin industry was growing rapidly in the late nineteenth centruy, Indonesia already had large export earnings from coffee and sugar, and tin by the first world war was a less important export earner than sugar, petroleum, tobacco and copra (Allen and Donnithorne, 1954, p. 291). Indonesia also developed rubber exports. Bolivia, however, was heavily dependent on tin for export earnings, following the decline of its silver exports.

Although analysis of the effects of tin mining on economic development is postponed until the case study in Chapter 5, it is worth commenting here that in Malaya government revenue from the export duty on tin was used to construct the core of the present-day rail and road system, and the main centres of population developed originally as tin mining towns. This infrastructure would later facilitate the spread of the rubbber industry. In the other producers, tin was in more remote areas.

NOTES

1. Nigeria is discussed at greater length in Chapter 3. Its annual production did not reach 1000 tons until 1911. South African and Australian production was larger, but then declined, while Nigerian importance grew.
2. The output is 'apparent' because there was considerable smuggling from the Dutch East Indies to the Straits (mainly Singapore) (Heidhues, 1992, ch. 2).
3. The material in this subsection on Malaya mainly derives from Thoburn (1977b, ch. 4), which cites the original source material, mainly British Colonial Office records. Some material from Thoburn (1981a, ch. 4) is also used in this and later subsections. On Thailand, Cushman (1991) and Falkus (1989) are the most useful sources.
4. Since the UK was the largest source of imports into Siam, the price of Siamese imports would probably have been influenced by the level of UK prices, hence the deflator for Figure 1.1 would be appropriate as a rough indicator of the real tin price facing Siam, as well as that facing British consumers. See the Appendix, and notes 6 and 8 to Chapter 1 for further discussion.
5. Smith (1980–2) says the very first application of dredging to tin mining was in Tasmania. The first dredge in Malaya was tried in Selangor in 1912 (Yip, 1969, p. 132).
6. See also Wong (1965, pp. 163–7 and 227–30) for further details, and Allen and Donnithorne (1954, pp. 158–9).
7. Heidhues' (1992) study of tin mining in Bangka (and Belitung) is the most useful source of reference on the development of the Indonesian tin industry from the earliest colonial times to the present; much of the information in this subsection derives from it, though other work (set out in Thoburn, 1981a) is also incorporated. See also the paper by Jackson (1969) in which the history of eighteenth century Bangka is researched from a rich variety of sources.
8. Other meanings of the term 'kongsi' are discussed at length in Heidhues (1992, pp. 37–9). In modern Chinese the term (written in Pinyin romanisation as *gongsi*) is that used for a company. The term is widely used in Malaysia too.
9. Heidhues (1991) says that prior to 1892 the Dutch colonial state already had a one-tenth share of the ownership and profits of the Billiton company.

10. This information on Billiton is from an account in Dutch of the company's early history by Mollema (1922); I am grateful to Elisabeth Onians for having translated it for me some time ago. Heidhues' (1991) paper on Belitung's history is also very useful, and I should like to thank K. W. Thee in Jakarta for bringing it to my attention before Heidhues' main work became available in print.

11. This section draws mainly from Klein (1965) and Hillman (1984).

12. Comments on other primary commodities, and on the development of colonial Malaya outside the tin industry, are based on Thoburn (1977b).

CHAPTER 3

The First World War and the Inter-war Years, 1918–1940

The first world war, unlike the second, caused little disruption to the world tin industry, but major structural changes took place in the industry during the interwar years. Western involvement in South-East Asian tin mining increased and consolidated on the basis of the new technologies introduced before the first world war. Vertical integration into smelting grew, and horizontal integration rose with the growth of new international tin companies. The tin market was highly volatile. There were severe falls in price after 1918 and again during the world depression starting in 1929, and attempts were made to control the tin market by international agreement. This chapter first looks at changes in the general pattern of production in the tin mining industry. It then examines the changes in industrial organisation which occurred, particularly with regard to the formation and influence of the two largest international tin groups – Anglo–Oriental/London Tin Corporation and Patino. Finally, there is an account of the Bandoeng Pool of 1921–5 and the international tin cartel of the 1930s and 1940s. The effects of the first world war are discussed in the individual sections where relevant.

CHANGES IN THE PATTERN OF TIN PRODUCTION

In 1918, as Table 3.1 shows, there were three large producers of tin – Malaya, Bolivia and the Dutch East Indies – who together accounted for almost 72 per cent of world output of tin-in-concentrates. The main new players in the inter-war period were Nigeria and, by the late 1930s, the Belgian Congo (Zaire). Siam gradually increased its output too, and produced a larger output than Nigeria in all of the inter-war years except 1926–9 (ITRDC, 1937). In 1929, the year of largest ever production prior to the onset of depression, Malaya, Bolivia and the Dutch East Indies accounted for almost 75 per cent of world production. This high degree of concentration of

TABLE 3.1   Production of tin-in-concentrates, by country, 1918–1940

(000 tons)	1918	1920	1925	1929	1933	1935	1940
Austria, Czechoslovakia and Germany	0.24	0.15	0.10	0.12	–	0.03	0.6
Portugal and Spain	0.20	0.19	0.70	0.89	0.62	0.96	2.0
UK	3.9	3.1	2.4	3.3	1.5	2.0	1.6
Nigeria	5.9	5.2	6.3	10.7	3.8	7.0	12.0
South Africa	1.4	1.5	1.1	1.2	0.54	0.62	0.59
Belgian Congo (Zaire)	0.11	0.33	1.1	0.97	2.2	6.5	12.3
USA	0.06	0.02	0.01	0.03	–	0.04	0.39
Bolivia	28.8	29.1	32.2	38.1	14.7	27.1	42.1
Burma	0.65	1.6	1.6	2.6	3.1	4.5	4.5
China	8.3	10.6	8.9	6.8	8.1	9.4	6.3
Indo-China	0.14	0.17	0.58	0.83	1.0	1.4	1.5
Japan	0.17	0.20	0.38	0.90	1.5	2.2	1.8
Malaya	40.1	36.9	48.1	69.4	24.9	46.0	83.0
Siam (Thailand)	8.8	6.2	6.8	9.9	10.3	9.8	17.1
Dutch East Indies (Indonesia)	19.8	21.6	31.3	35.7	14.4	24.7	41.3
Australia	4.7	5.2	3.0	2.2	2.8	3.1	3.9
World	123.9	122.4	145.0	192.2	90.8	146.8	235.0

Sources and Notes
1. From ITRDC (1937) for 1918 to 1935. 1940, and statistics for Burma, are from ITSG (1949). 1940 figure for Spain and Portugal also includes Italy, 1940 figure for USA includes all North and Central America. There are some slight differences in figures between the two publications.
2. Countries not producing over 100 tons in any year 1918–40 have been excluded.

production was significant for the formation of the tin cartel in the 1930s, which is discussed in the final section. China, though a large producer, largely remained outside international markets, and her tin industry in Yunnan was controlled by warlords. British and Australian production declined through much of the inter-war period, 'though Australian output rose slightly as other countries restricted their tin production in the 1930s under the international tin cartel. South African production remained very low, and declined in the 1930s to well under 1000 tons.

In Indonesia, the Dutch East Indies, tin remained a government monopoly on Bangka island, and the Dutch Billiton company controlled production on the country's other two mining islands. There was no (non-Dutch) foreign investment. In Malaya there was a rapid expansion in the number and size of Western companies as dredging grew in importance, and there were similar developments in Siam. Western involvement grew also in Nigeria and Burma. In Bolivia the industry progressively became monopolised by the three main groups (Patino, Hochschild and Aramayo), with a fringe of medium and smaller mines.

South-East Asia

In the Dutch East Indies, the colonial administration which also managed the mines on Bangka island had been replaced in 1913 by a state mining organisation BTW (Bangka Tinwinning). In 1918, nearly four-fifths of the miners who returned to China did not have savings to take back with them, worsening Bangka's already bad reputation as a place to work; one commentator estimated that of all wages paid during the period 1912–18, 95 per cent were retained in Bangka island. Labour recruitment remained a problem until the 1929 depression. A penal sanction against mineworkers leaving their employment remained in force up to the end of the 1930s, though by then it was starting to die out. At the end of the 1930s most mining labour was still ethnically Chinese, and still mostly immigrant. However, ethnic Indonesians proved willing to work with the new techniques – gravel pumping and dredging, whose use spread rapidly from the 1920s. *Kongsis* as a form of organisation were abolished on Bangka from about 1920, and mine operators were made employees of BTW, which continued to provide machinery, materials and food as it had done under the previous system. Labour conditions on Bangka did not greatly improve as a result of this change, however (Heidhues, 1992, ch. 5).

In Belitung after the First World war, with changes in mining technique away from labour-intensive open-cast mining, more mines started to be run directly by the Billiton Company, with workers paid on hourly rates. On dredges workers worked directly for the Company. At its peak in 1921–2, the mining labour force on Belitung was about 20,000 (there were 21,000 on Bangka at that time), mostly ethnic Chinese. From the 1929 depression, thousands of displaced miners were repatriated to China at the Company's expense. In 1924 the Billiton Company was reorganised into the Billiton Joint Mining Company, in which the colonial government held five-eighths, with three-eighths held by the Billiton Company of the Netherlands (Heidhues, 1991).

By the late 1920s all Belitung's tin was being smelted by the Straits Trading Company in Singapore, and there were a few smelters on Bangka, compared to over a hundred in 1900. By the 1930s only three smelters were left on Bangka, and Dutch East Indies tin was smelted mainly overseas (Heidhues, 1992, p. 128).

In Malaya, the foundations for a rapid expansion of Western-owned mining had been laid before the first world war, with the development of

TABLE 3.2   Production of tin in Malaya, by mining method, 1928–39 (percentages
             of total output)

	1928	1930	1935	1939
Dredging	30	38	45	48
Gravel Pumping	45	43	41	36
Hydraulicing	8	7	5	6
Open Cast	6	4	2	4
Underground	6	5	4	4
Miscellaneous	2	1	1	–
Dulang Washing	2	1	2	2

Sources and Notes
1. From Thoburn (1977b, p. 87).
2. 1928 is first year for which sectoral output figures are available.
3. Figures are for Federated Malay States only

dredging. There was a boom in Western mining investment in the late
1920s, and by 1928, as Table 3.2 shows, nearly a third of Malayan output
was being produced by dredges, all Western-owned. By the same time,
open-cast mining had declined as a technique, and was replaced among
ethnic Chinese miners by gravel pumping. The relatively low labour intensity of dredging meant that though by 1939 it was producing nearly half the
country's output, this generated only about a quarter of Malaya's mining
employment (see Table 3.3).

In Siam, dredging, which had been producing over 30 per cent of the
country's tin output as early as 1915, by 1928 was producing over 40 per
cent (ITRDC, 1937, p. 39). By the second world war, dredging's share of the
country's tin output was over 60 per cent (Ingram, 1971, p. 100). Siam sent
its ore to the Straits Settlements for smelting. Both the Western and the
ethnic Chinese sectors of the Siamese tin industry continued to have strong
Malayan connections. Unlike Malaya, but like the NEI, much of Siam's
dredging was offshore.

Bolivia

For most of the inter-war years until the late 1930s, Bolivia was the world's
second largest producer of tin, ahead of the Dutch East Indies. During the
second world war from 1942, when the South-East Asian producers were
occupied by (or, in the case of Thailand, allied to) the Japanese, Bolivia
became the world's largest producer.

An interesting feature of the Bolivian case in this chapter is that Bolivia
was the starting point for the development of an influential multinational,
the organisation founded by the Bolivian Simon Patino and headquartered
in the United States after 1924. In Jones' (1925) account of the Bolivian tin

TABLE 3.3   Employment in mining in Malaya, 1915–39, by mining method (percentages of total mining employment)

	Dredging	Gravel pumping	Hydraulicing	Open-cast	Underground
1915	1	20	–		72 7
1920	3	39	–	47	11
1925	7	46	–	27	7
1930	15	51	11	15	7
1935	17	49	8	18	8
1939	23	50	5	16	7

Sources and Notes
1. From Thoburn (1977b, p. 88).
2. Figures refer to Federated Malay States only.
3. Gravel pumping figure for 1915–25 includes hydraulicing.
4. Underground mining is slightly overstated since it includes a small amount of
   coal mining employment, and open cast includes some iron mining. Nevertheless, the employment figures are domininated by tin.
5. Dulang washers are not included in total employment.

industry in the 1920s many Chilean and some British and French companies were listed; foreign companies apparently accounted for two-thirds of
the capital in the industry (Hennart, 1986b, p. 256). Patino in fact was the
instrument whereby this foreign control was greatly reduced. During the
boom of the mid-1920s Patino expanded and acquired control of other
mines, so that by 1929 his group controlled some 60 per cent of Bolivian
tin. Two other groups were of importance: Hochschild, who had started as
an ore dealer for small mines in 1921 and by 1929 was controlling a large
proportion of their supplies; and the Aramayo company (Hillman, 1988a, p.
86–7). At the outbreak of the second world war Patino controlled 45 per
cent of Bolivian tin output, Hochschild 23 per cent and Aramayo 7 per cent
(Hillman, 1990a, pp. 291–2). These three groups continued to dominate the
industry until their nationalisation in 1952.

Patino's growth had originated in his finding an exceptionally rich
deposit in his first mine, which by 1910 was producing some 10 per cent of
world production. A series of acquisitions of French, British- and Chilean-
owned companies, and railroad and banking interests, led to the formation
of a group – Patino Mines and Enterprises Incorporated – incorporated in
the USA. The decision to locate in the USA is ascribed by Klein (1965) to a
wish to tap US sources of capital, and Patino had indeed used credit facilities skilfully to build up his group.[1] It also greatly lessened, of course, the
Bolivian government's control over the company's activities, and the reor-
ganisation accelerated a process of acquisition which was to lead to interests
in Asia, Europe and North America.

Besides the Patino interests in Bolivian tin, banking and railways, one very important diversification had already taken place before 1924 – the group had acquired control of Europe's largest tin smelter, Williams, Harvey in Liverpool, built in 1909. This had been done in partnership with the National Lead Company of the US, at the time the world's second largest consumer of tin, which it used for solder, babbitt (alloys used for lining bearings) and other bearing metals (Hennart, 1986b, p. 267). Their threat in 1916 to build their own smelter had allowed them to buy half of Williams Harvey's equity, and they acquired the rest soon after; the first half being acquired two-thirds to a third and the second half each. National Lead was also an active participant in the takeover of the largest mine in Bolivia, previously owned by Chileans. Patino also had acquired previously an interest in the German tin smelter to which his ore had been sent before the first world war blockade on Germany made him switch to the UK for smelting (Hennart, 1986b, p. 231).

The National Lead–Patino combination represents a grouping of consuming and mining interests unusual in the tin industry, even today.

Nigeria[2]

Nigeria had been producing tin for local use since pre-colonial times, with local smelting and tinsmithing, but there was minimal sale outside the country until the twentieth century. In the early colonial period before the first world war, the Jos Plateau region (the centre of Nigerian tin mining) was brought under control by harsh military means. Legislation was passed banning local smelting and, in effect, restricting Africans from operating mines. As the industry developed under Western control, ore was shipped to the UK for smelting (and this continued until 1961). Given that Africans had little opportunity to participate except as labourers, the interplay between ownership and technique, which was so important in Malaya and Siam in establishing a local (albeit immigrant Chinese) presence, was largely absent. Western companies themselves used a variety of mining techniques.

Prior to the first world war, there had been much London-based speculation in Nigerian mining investment, and foreign companies such as Consolidated Gold Fields from South Africa had shown interest in Nigerian tin. Most of the actual mining, however, was done using traditional local techniques. The low tin price in the 1921 depression brought about the collapse of some companies, including the most important, and the consolidation of others. As the world tin industry recovered in the mid-1920s mechanisation began, with Western capital. Hydraulicing was used too, and a Consolidated Goldfields-owned company used a bucket dredge. From 1926 the Anglo–Oriental financial group (see next section) started to become a force in the world tin industry, and quickly moved into Nigerian tin. The Anglo–Oriental company Associated Tin Mines of Nigeria was formed, which made a series of acquisitions. By 1929 Anglo-Oriental controlled 41 per cent of Nigerian tin production, and also owned highly profitable hydro-electric power generating facilities. There also had been a growth of small scale mining in the 1920s, often started by British mining engineers, although from 1926 Africans were also able to participate for the first time. These small miners depended on large companies for power and sometimes for finance, and tended to use simpler, often pre-colonial techniques.

During the depression from 1929, which also saw the first appearance of African mine-leaseholders, there were large reductions in mining employment (from the inter-war peak of nearly 40,000 to a low of under 15,000 in 1933) and great cuts in wages. Tin control (see later) was strictly enforced. Companies tended to return to more labour-intensive methods when prices were low. Hand methods were generally used to work the richest deposits, and many deposits were not suitable for large-scale mechanisation by dredges along South-East Asian lines. Though more labour-intensive, the share of wage payments in total output in Nigeria appears to have been no larger than that in Malaya.[3] However, the degree of mechanisation was sufficient to bring into being a skilled labour force, and small-scale engineering developed to service the mines.

By 1940, Freund (1981, p. 127) argues that the best deposits in Nigeria were gone, and although production was increased during the early 1940s to help the Allied war effort, the pattern thereafter would be one of decline.

Other Tin Producers

In China, which in the 1920s was the fourth largest producer, mining remained under local control and attempts by Western firms even to buy concentrates for smelting were firmly resisted. Over 90 per cent of output at that time came from Yunnan province. Tin was also produced in Guangxi, Hunan and Hainan island (ITRDC, 1937, p. 29). The methods then used, as described by Jones (1925, pp. 240–5), sound similar to the older open-cast methods used in Malaya in the 1870s, and the deposits then being worked are described as alluvial. Tin for domestic use was smelted in Shanghai.

Dredging seems to have spread rapidly among alluvial tin producers, and Jones' survey of the world tin industry published in 1925 records dredges in operation in Burma, Nigeria and Australia. In Burma, then part of the 'British India', Jones (1925) notes a number of Western companies in operation, with gravel pumping also being quite widely used. At Tavoy, in the south of the country, a suction dredge operation was pioneered by a British company. Similarly in Indo-China some Western involvement was recorded.

In both countries the first world war had provided a stimulus to mining because of the demand for tungsten for armaments, tin and wolframite (tungsten) being found together in many areas in the two countries.

In Australia, however, as in Nigeria and the Dutch East Indies, the interesting interplay between dredging and the growth of foreign companies was lacking. In Australia the industry was developed mainly by Australian companies. In the Congo, even after the industry was developed rapidly with Belgian capital in the 1930s (Fox, 1974, pp. 52–3) dredges were unimportant, the deposits either being very small or very near the surface and amenable to simple open-cast methods.

In Brazil, tin had been discovered in 1903, at the Campinas mine in the southern state of Rio Grande do Sul. During the interwar years there was only the most minor, intermittent production of tin. Jones (1925) did not mention Brazil as a producer at all. The small size of production, until the discovery of deposits in Minas Gerais state in 1943, can be judged from the fact that when Brazilian production was stimulated by wartime demand in the early 1940s it remained a very minor producer and a net importer of tin (ITC, 1960, p. 94). In none of the war years did production from areas other than Minas Gerais total as much as 20 tons (ITSG, 1949, p. 113).

CHANGES IN INDUSTRIAL STRUCTURE

During the late 1920s and the 1930s a process of consolidation occurred, bringing about a considerable degree of horizontal integration and strengthening the links between mines and smelters. The two main developments were the formation in 1925 of what was to become the London Tin Corporation, and the establishment of Patino's Consolidated Tin Smelters in 1929.

Until the 1920s much of the British involvement in Malayan tin – and most 'Western' involvement was British – had consisted of large numbers of 'free-standing' companies, developed either by expatriates or directly from Britain, but without tin interests elsewhere (van Helten and Jones, 1989, p. 166); in other words, they were not 'multinationals' in today's sense. The Cornish tin interests associated with Osborne and Chappel were exceptions to this rule, but even these represented more of a migration of capital from Britain than genuine 'multinationalisation'. The Malayan tin companies played no role in the development of the tin industry in Siam before the first world war, although the agency house Guthries from British Malaya did have some involvement. The upsurge of investment in Malayan tin in the late 1920s, as tin prices recovered from the recession after the first world war, was associated with increasing concentration in the industry, in which the Anglo-Oriental played a large part. The 1930s depression also allowed Anglo-Oriental to buy up tin companies in difficulties as share prices fell. Links in the tin industry between Malaya and Siam strengthened in the inter-war period. The Renong Dredge Company in Siam had properties in Malaya to which it transferred dredges in the 1920s as the Thai deposits reached exhaustion (Jones, 1925, p. 217). Kamunting Tin, one of the first dredging companies in Malaya, operated in Siam too, and continued to do so until well into well after the second world war.

Anglo–Oriental had been started in 1925 by the financier John Howeson, later to be jailed for issuing a false prospectus in connection with an attempt to manipulate another commodity market – pepper (Burke, 1990, pp. 48-52). The London Tin Corporation was set up by Howeson in 1929 following a series of acquisitions in Africa (Freund, 1981, p. 118). By 1937 the group, as the London Tin Corporation (LTC) operating through its subsidiary Anglo–Oriental (Malaya), controlled twenty companies in Malaya producing about a third of the country's output and roughly half that of the foreign sector (Allen and Donnithorne, 1954, p. 157). Companies in Thailand came under LTC control too, such as Tongkah Harbour in 1934 (Falkus, 1989, p. 154), and also in Burma, while the LTC-controlled Associated Tin Mines of Nigeria which accounted for half of that country's output. The LTC did not normally have majority ownership, but control could be exercised since other shareholdings were dispersed, besides which Anglo–Oriental acted as managers. Control was operated through a combination of management and interlocking directorships, a pattern that persisted into the 1970's.[4]

In 1920, over 60 per cent of world tin ore output was being smelted in the non-industrialised producing countries: 41 per cent of world smelting was carried out in Malaya, 11 per cent in the Dutch East Indies and 9 per cent in China. Australia (3 per cent) also smelted part of its own output, while most of the rest of the world smelting was carried out in the UK (20 per cent) and the USA (13 per cent). American participation in world smelting, which developed during the first world war, was temporary however. US smelter output peaked at over 15,000 tons in 1920, but declined thereafter as smelters were hit by recession, and there was no reported output after 1924 (ITRDC, 1937, p. 25). Only minimal amounts of domestic mine production in Bolivia and Thailand were locally smelted (Jones, 1925, p. 25). In Malaya the smelting was divided roughly equally between the Straits Trading Company, founded by Europeans as an independent venture in the late nineteenth century, and Eastern Smelting, which was set up by local Chinese, but taken over by British interests before the First World war, as Chapter 2 has shown.

Consolidated Tin Smelters (CTS) was formed by Patino, following his group's acquiring control of the Williams, Harvey smelter in England[5] which then smelted much of Bolivian production. Besides two other smaller British smelting companies, control of Eastern Smelting in Malaya was secured[6] and CTS also held 'substantial' interests in the Dutch smelter at Arnhem, which by the early 1930s was smelting most Indonesian output (which previously had been smelted in Malaya) (Allen and Donnithorne, 1954, p. 160). The interest in the German smelter was held jointly with National Lead, and in the Arnhem smelter with Billiton (Hillman, 1988a, p. 86).

Of the major smelters serving the international economy, there thus remained outside CTS only the Straits Trading Company in Malaya. The existence of the Straits Trading Company as a large independent smelter, which also had some holdings in mines in the Osborne and Chappel group, meant that independent miners in Malaya were not subject to significant monopolistic pressure and most marketing remained independent of the large groups. Patino's acquisition of smelters was clearly an attempt to secure his market position, and virtually no integration forwards into tin using industries occurred, nor did tin users themselves acquire interests in mining or smelting other than National Lead's cooperations with Patino.

Between the Anglo–Oriental and Patino groups there were a number of cross holdings. Anglo–Oriental, which had built a smelter in Britain in 1928, invested in Consolidated Tin Smelters (Freund, 1981, pp. 116–17). Geddes (1972, p. 226–8) mentions that Patino actively engaged in buying large quantities of shares in major Malayan tin companies, and cites British press reports to the effect that Patino owned 80 per cent of the shares in certain large LTC operating companies. The LTC–CTS combination was a major influence for the establishment of tin production control schemes in the 1930s (see later).

The inter-war structure of the world tin industry was also one which gave considerable economic strength to Britain as a colonial power, particularly in relation to the United States.[7] An American preference for Straits tin meant that much of Malayan output was sold to the United States. However, the export tax on tin concentrates from Malaya, dating from before the first world war, prevented other countries from securing Malayan ore for smelting, and indeed was a measure designed to frustrate past attempts by the Americans to take control of the Malayan tin trade (Wong, 1965, pp. 229–30). The substantial smelting capacity in Britain was used to smelt high grade Bolivian ores, mainly from Patino's mines, and this was the basis of Patino's interest in Williams, Harvey. The Bolivian ore was mixed with alluvial concentrates from Nigeria, the export of which to other countries was also prevented by an export tax (imposed in 1915 – see Freund, 1981, p. 117). The remaining high-grade concentrates from Bolivia were insufficient to support a competitor smelter. Medium-grade concentrates were treated at the Arnhem smelter, and low grades in Germany and at Capper Pass smelter in Britain. The United States, with no significant domestic production, depended heavily on British-controlled supplies, and would further resent the establishment of the tin cartel in the 1930s to raise price, particularly after demand had started to recover after the depression (Yip, 1969, p. 284). Wartime conflicts over tin policy between Britain and the United States, as the US tried to develop its smelter capacity on the basis of Bolivian ores in part attracted away from British smelters, springs from an attempt to change this balance of power. The stockpile of tin, which the Americans started to accumulate during the early phases of the second world war, would at last give it market power and delay and dilute any attempts at post-war cartelisation.

TIN CONTROL

The Bandoeng Pool

During the latter years of the first world war there had been a rapid rise in the tin price as shortages developed as a result of shipping difficulties, while at the same time stocks accumulated in producer countries. This growth in stocks, when combined with a world slump in 1921, reduced the average annual price in real terms by more than 50 per cent between 1918 and 1921 (see Figure 1.1). In December 1920 the government of the Federated Malay States announced a minimum price at which it would buy and stock Malayan tin. This support buying was maintained until February 1921, when the FMS government agreed with the government of the Dutch East Indies to form a pool – the Bandoeng Pool – to withhold tin from the market. At the time, the two countries together produced roughly half of the world's tin. From 1921 some 19,000 tons were withheld from the market – equivalent to about 15 per cent of 1920 world production. The tin was held until June 1923, and as prices strengthened, the tin was gradually released. The stock was exhausted in 1925, by which time world consumption had almost doubled (to 150,000 tons) compared to the trough of the slump in 1921 (when consumption was 80,000 tons). Not only was the pool successful in reducing the price instability of the tin market, but the FMS government was able to announce a substantial profit from the operation. (Yip, 1969, pp. 155–60).

The International Tin Cartel

There was a rapid expansion of tin output from the early 1920s as prices recovered.[8] In Malaya, the largest producer, output increased to 1929 by over a third compared to its peak of over 50,000 tons before the first world war. Over the period 1920–9 Malayan output almost doubled. As we've seen, this expansion of output was associated with a growth in share of the Western-run dredging sector. The Dutch East Indies apparently had a conservative depletion policy, maintained by tight control over the industry, under which it mined high grades in times of low price and low grades in times of high price in order to maintain stability in its income flow. This is in contrast to Malaya, and Siam, where Western firms were allowed to expand in conditions of competition (Hillman, 1988b, pp. 241–2). Nevertheless, from 1920 to 1929 Dutch East Indies' output rose by about two-thirds. The fall in price from the 1926–7 (annual average) peak is shown in Figure 1.1, and as world depression deepened the price collapsed in real terms to a level almost as low as that immediately after the first world war, reaching its trough in 1931–2.

Tin had been withheld from the market by Anglo–Oriental and Patino in 1928 (Fox, 1974, p. 122), but their withholding of stocks was insufficient to maintain prices. By mid-1929, with falling prices and further accumulation of stocks, a voluntary Tin Producers Association (TPA) was proposed (in a letter to *The Times* of London), and Simon Patino became its honourary president in January 1930. The TPA would stockpile tin concentrates at the smelters, who would restrict their output of metal. The TPA also engaged in some piecemeal output limitation, but had insufficient control over supply for this to be effective in the face of falling world consumption. This failure led to a campaign by the TPA participants for formal tin restriction. The tin cartel started operation in 1931.[9] Three successive agreements ran 1931–3, 1934–6, 1936–41, and a fourth was negotiated to run 1942–6 in the hope of being able to handle an anticipated postwar recession.

In the light of post-war attempts to control the tin market, several features of inter-war tin control are of interest. Some members were able to secure more favourable terms in the successive negotiations for each agreement than were others; the inter-war agreements pioneered the use not only of output controls, but also the use of buffer pools and stocks to influence the market; and there were conflicts between the Americans and the members of the tin cartel.

When the cartel was being negotiated, conditions for cartelisation appeared favourable, in the sense that the three largest producers (Malaya, Bolivia and the Dutch East Indies) in 1929 had accounted for nearly 75 per cent of world output; this amount rose to 80 per cent with Nigeria and 85 per cent with Siam as well. There were divisions both between and within producing countries, however. Cartelisation always involves a free rider problem. If particular producers feel the others will go ahead regardless, the outsiders can enjoy the benefits of higher prices without the pain of restriction. This is particularly a temptation for small producers, who do not feel their participation would be crucial to the cartel's formation. Among major players there may be differences in bargaining power. Low cost producers will feel they can withstand low prices in the free market for longer while they hold out for agreement on better terms. Countries with tight control over their industries will be subject to fewer pressures from divergent interest groups. The fact that mining companies may control production in several countries may cause divergences between the interests of the mining companies and the various national governments. In tin in the inter-war period, colonial powers were involved too. Consumers also have an obvious interest in prices. Though they may be in favour of price stability, they are unlikely to want it in the context of attempts to raise trend prices by long-term production restriction.

Negotiations were held between the Dutch, Anglo–Oriental, Malaya and Bolivia in November 1930. The Dutch East Indies, Malaya, Bolivia and Nigeria became the first signatories, in 1931. Later in 1931 Siam also joined the cartel. The Belgian Congo joined in 1933 and remained, having been given a standard tonnage (see below) which increased each year. Portugal, French Indo-China and the producers association of Cornwall followed and joined the second agreement. Portugal and the Cornish producers did not join the third, however. The cartel members apparently did not think it was worth approaching the Yunnan warlords who controlled most of China's tin industry, feeling that in any case their output levels could not have been monitored.[10] Strong support for primary commodity regulation was given by the British Colonial Office, which regarded it as a means of alleviating the effects of depression on colonial economies,[11] and the government of the FMS also was keen to have tin restriction. Starting with the second agreement, consumer representatives were invited to attend committee meetings, and Fox (1974, pp. 160 and 178) records a delegate who represented British tinplate interests expressing strong views about the tin price.

The Dutch were operating an industry under tight control – directly under the control of the state in the case of Bangka Tinwinning, and indirectly on Belitung and Singkep islands through the Billiton company, which was five-eighths state-owned. Within Malaya, the Anglo–Oriental interest led by John Howeson was opposed by the older Western mining interests in Malaya, who had sprung from strong Cornish connections. There were suspicions that Anglo–Oriental as a financial group wanted to secure control over tin stocks to increase their power in the market. Also, other Malayan producers felt that the Anglo–Oriental group's expansion of production in the late 1920s would mean excessive output reduction by existing mines in order to accommodate Anglo–Oriental. Nevertheless the older interests supported the idea of restriction, though they wanted it shorter and sharper than the two years initially proposed. At the beginning of restriction, there were strong fears on the part of the FMS government about the effects of sharp cutbacks on the small mines owned by Chinese in Malaya. Chinese miners in Malaya had voted heavily for restriction in a referendum in 1931, but this was before the terms of restriction were known. Yip (1969, p. 275) argues that had they known the size of the cuts and the length of restriction, their support would not have been assured.

Production was regulated by means of output quotas established in relation to a 'standard tonnage', with some minimum output guarantees for producers. In Nigeria special minimum output provisions were given for small-scale miners. Similar provisions were not negotiated for Malaya, though they could have helped protect the ethnic Chinese sector. It seems that in the negotiations for the first agreement, Bolivia was able to secure particularly favourable terms, since the standard tonnages were based on 1929 outputs. Parts of Bolivian capacity had been wiped out in 1930 as price fell and the Bolivian exchange rate was kept high as part of domestic policy to secure cheap food and manufactured imports. As a result, Bolivia's standard tonnage was in fact in excess of her capacity at the start of restriction. Yip (1969, ch. 11) argues that 1929 output greatly understated Malayan capacity, since, in addition to the 105 dredges in operation, there were ten under construction in 1929 and another eight on order. This, together with the ignoring of Malayan output outside the Federated Malay States and rises in smaller-scale mining capacity in 1929 meant Malayan capacity was about 35 per cent higher than that represented by its standard tonnage.

When small Bolivian mines overproduced in relation to quota as the price rose again towards the end of the first agreement, Bolivia was able to secure larger standard tonnage under the second agreement at the expense of the Dutch. The Siamese government took a hard line after the country's 1933 revolution, and was also able to secure favourable terms – particularly in terms of its tonnage under the third agreement.

Hard bargaining took place too over the withdrawing of excess tin stocks into a buffer pool during the first agreement, and with regard to setting up a buffer stock from new tin production under later agreements once the immediate problem of excess stocks and production had been solved by production restriction. The Labour Government In Britain apparently was unable, for reasons of financial orthodoxy, to permit public money from Malaya to be used to finance a buffer pool under the first agreement, despite the earlier success of the Bandoeng Pool (Hillman, 1988a, p. 94). In consequence, three syndicates were set up through negotiations with the cartel – one British, based on Anglo–Oriental and some other British interests; one comprised of Billiton and the Dutch East Indies government; and a Bolivian one centred on Patino. Patino reneged on this agreement, although he set up a private pool of his own a year later. The British and Dutch syndicates went ahead with a pool, which withdrew 21,000 tons from the market, albeit with an agreement from the Bolivian government that it would remain a cartel member until the excess stocks had been disposed of. However, Patino's failure to participate in the pool does appear to have weakened the cooperation widely assumed in the literature to have existed between Patino and Anglo–Oriental in the inter-war period, as well as making the Dutch very suspicious of Bolivian intentions.

The gains of Siam and Bolivia in relation to Malaya and the Dutch East Indies were secured despite an apparently strong Dutch bargaining position. The Dutch East Indies was a low-cost producer with tight control over its tin industry. Siam's strength lay in its potentiality as a free rider, as also did that of smaller producers like the Congo. Then a relatively small producer and dependent on tin only for about 10 per cent of its export earnings, Siam could in principle wait for the other countries, to whom the tin industry was more important, to reach agreement. In fact, despite not been under the rule of any colonial power, Siam was constrained to some extent in its willingness to act as a free rider by its fear of other interests being damaged if it behaved unreasonably on tin. Its internal politics after the military coup in 1932 caused it to act more toughly, although the Siamese tin industry itself was apparently more hardline than the government throughout. Matters were also complicated by the fact that various Western interests had mines both in Siam and Malaya, as presumably did some Chinese miners too. Although the Dutch at times appeared ready to threaten economic warfare in the tin market against Siam, the Malayans (especially in the negotiations for the third agreement) feared the consequences for their own industry and were willing to concede to Siam. There were fears on the part of the British that if Siam were not accommodated under the third agreement, it might be tempted to ship ore to the USA to help the US develop tin smelting.

Bolivia's position, Hillman (1988a) argues, citing some earlier sources, paradoxically was strong because of its heavy dependence on tin for export earnings. In his view, although Bolivia was said even in the inter-war period to be 'high cost', its tin export earnings had such a strong potential influence on the country's exchange rate that a collapse in the tin price would put great pressure on the real exchange rate to depreciate. Although it would reduce import capacity, the depreciation would make tin exports competitive again at world prices. In comparison, a country such as Siam presumably would have its real exchange rate determined more by flows of traded goods other than by tin. This argument is overstated to the extent that the disabsorption (cuts in real consumption, investment and government expenditure) normally necessary if devaluation is to improve the trade balance may be difficult to bring about.[12] Also, if the tin price collapse occurs when other primary commodity prices are collapsing (as was the case in the 1929–31 depression – and certainly so for rubber, Malaya's largest export and the Dutch East Indies' too in the mid-1920s) then the exchange rates of the South-East Asian and African tin producers would have also depreciated, cancelling out the effect. However, in the event that, say, the rubber price was supported by restriction (as it was), but the tin restriction scheme was abandoned, the argument would have some force. At least, at crucial junctures, the Dutch believed that the Bolivians could not be pushed out of the market by a period of economic warfare in which the tin price was allowed to collapse (Hillman, 1988a, p. 98). Bolivia's negotiating position was also strengthened by the fact that it had not joined the buffer pools associated with the first agreement, so failure to agree (in this case with regard to the 1934–6 scheme) would have left British and Dutch interests with losses on their tin pools.

The buffer pools set by private syndicates in parallel with the first agreement lasted from 1931 to 1934, and in 1932 they held half the world's visible supply of tin metal (Fox, 1974, p. 152). As the pools began to sell tin in 1933, proposals for a buffer stock to be set up directly under the auspices of the International Tin Committee was proposed by Howeson of Anglo–Oriental, to be generated by special production quotas. In the event, the buffer stock under the second agreement was very short-lived, and met much opposition – in Malaya there was resentment about the pools' profits, which were made while Malaya had suffered particularly heavy restriction. The Dutch were keen to set up buffer stocks to stabilise the market under the second and third agreements. This was resisted by the Bolivians for the second agreement since they were underproducing in relation to quota and wished to raise production. The buffer stock set up under the third agreement came into effect in 1938, with 10,000 tons of tin, soon increased to 15,000. It had a brief to intervene in the market when prices moved a certain percentage away a stated pivotal price. Fox argues it had an important stabilising effect on price as demand rose in 1939 prior to the outbreak of war. Once war was declared the buffer stock soon exhausted its tin, and by early 1940 it held only cash.

Table 3.1 shows the severity of the fall in output brought about by the recession and subsequent restriction. Each of the largest producers – Malaya, Bolivia, the Dutch East Indies and Nigeria – was producing in 1933 an amount equal to some 40 per cent or less of their peak outputs in 1929. Generally the major producers in the cartel restricted output proportionately more than both smaller members and, of course, more than the non-members. In the first scheme the degree of restriction was sharp – the quotas were equivalent to 69.5 per cent of standard tonnage in 1931, 44.4 per cent in 1932, and 33.3 per cent in 1933. During the course of the first agreement, output of non-member countries rose almost by half in absolute terms, and their share of world tin output increased from 10 per cent to 25 per cent of world output (Yip, 1969, ch. 11). Thereafter, the efforts to secure a higher membership succeeded, but generally there were more generous allocations for new members than for founder members

By 1940, a year in the second half of which there was unrestricted production, the combined shares of Malaya, Bolivia, the Dutch East Indies fell to 71 per cent, and 76 per cent including Nigeria, compared to 75 per cent and 80 per cent respectively in 1929. In the case of Bolivia, the depression was a major factor in the early loss of output. For much of the period of restriction Bolivia was an underproducer in relation to its quota. In the late 1930s until the end of the rearmament boom in 1938, Bolivia underproduced in consequence of the removal of labour from the mines to fight in the Chaco war against Paraguay.

The fourth scheme, negotiated by 1941, did not come into effect. Malaya and the Dutch East Indies were invaded by the Japanese, and Thailand allied itself to Japan. Conditions were changed greatly by the beginning of tin stockpiling by the United States in anticipation of wartime shortages. Negotiations between the US and the cartel resulted in agreement to remove production restrictions in mid-1940 in exchange for some guarantees about the rate at which stockpiled tin would be released by the Americans after the war. Bolivia continued to behave opportunistically in relation to the cartel, and increased production to supply the US with tin for stockpiling. Hillman (1988a, pp. 108–10) concludes that Bolivia gained most from, and contributed least to, the tin restriction schemes. For example, Bolivian marginal mines were kept going in the 1930s, and were then able to expand output during the shortages of the second world war. High tin prices enabled the state in Bolivia to take a higher tax revenue from the industry than it could prior to restriction. The influence of Patino on the world tin industry through his control of a large proportion of international smelting capacity did not give him as much power as is often thought, the supply of tin ore was well-controlled by other cartel members, particularly the Dutch.

Bolivia's continued opportunism did much to destroy the basis of trust on which effective international cooperation could be conducted. However, it was American hostility to cartels, and American power backed by its stockpile, which made the cartel impossible to resuscitate after the war. The cartel formally came to an end in 1946.

OVERVIEW: 1914–1940

The inter-war years saw little basic change in the geographical location of tin production, with Malaya for most years being the largest producer. Nigeria and the Belgian Congo (Zaire) established themselves as minor producers. From the time of the first world war, the United States became the largest consumer of tin, of which the most important use was tinplate, taking over that position from the UK. The US not only had no significant production of its own, but American companies had little involvement in tin production or smelting. Britain defended its colonial tin interests. In the recession of the 1920s and again in the great depression of the 1930s, tin output was restricted by international agreement, with the governments of Britain and the Dutch East Indies providing strong support to their tin interests. America had little control or influence over tin supply in the face of the powerful international tin cartel backed by Britain. Tin of course was not alone in its difficulties during the depression; most other commodity prices collapsed too. In a similar fashion to tin, Britain introduced a rubber control scheme in Malaya and Ceylon.

The 1920s saw a great expansion in foreign investment in tin, especially by British interests, though the production of many other metals expanded even more. Whereas other metals fell under the control of diversified international mining companies, the 'multinationals' in tin, though powerful in their field, tended to be specialised on tin-mining and smelting. The most important grouping was the London Tin Corporation, with interests in Malaya, Siam, Nigeria and Burma. The LTC had close links with the multinational owned by the Bolivian Patino, which had extensive smelting interests. The LTC/Patino grouping exerted an important influence on the provsions of the international tin cartel, which may have worked against the interest of other mining firms, especially the local sector in Malaya. There was very little involvement in mining or smelting by tin-using companies, whose main interests (in the case of tinplate) were linked to steel.

To anticipate the discussion of development effects in Chapter 5, there is little quantitative information on any producer for this period except Malaya. In inter-war Malaya, export duty in most years took around 15 per cent of the value of gross output of tin, providing the colonial government with an important source of revenue. For tin-mining as a whole in the inter-war period (in the absence of sufficiently detailed sectoral data), identifiable direct costs and export duty usually comprised less than half the value of output. For foreign companies the rest could have been repatriated as profits and to meet capital charges. In consequence, retained value in foreign-owned mines will have been much less than in the post-war period when taxation of profits was introduced. Gravel pump mining paid a higher percentage of its gross revenue out as wages (*c.* 20 per cent) than did dredging (*c.* 8 per cent), though the figures varied with the tin price and export control (Thoburn, 1977b, ch. 5).

NOTES

1. See especially Geddes' (1972) biography of Patino, chs 10, 12 and 16.
2. This account is mainly based on Freund (1981).
3. Freund (1981, p. 126) gives data for 1942 showing wage payments were 15.4 per cent of the value of tin output. Figures I have

assembled for the Federated Malay States in the 1920s and 1930s suggest a similar order of magnitude, though the year to year variation is considerable, and generally *higher* in years of high tin prices (Thoburn, 1977b, pp. 108–9). The average annual real tin price in 1942 was a little lower than in the mid- and late-1930s (see Figure 1.1).

4. For an exhaustive account of ownership and control in Malaya in the early post-war period, much of it inherited from the inter-war years, see Putucheary (1960).

5. According to Klein (1965, p. 19) when CTS was formed Patino bought out National Lead's holding in Williams, Harvey.

6. See Geddes (1972, ch. 18) for further details.

7. This paragraph draws on Hillman (1990a), who provides a penetrating study of conflicting Anglo-American interests in tin, played out through their respective relations with Bolivia during the second world war.

8. As Fox (1974, pp. 120–1) notes, however, the increase in the production of many other metals was even greater; the tin expansion was in no way exceptional.

9. The research of John Hillman (1984, 1988b, 1990b) has provided very useful analysis and detail about the formation and workings of the international tin cartel, and the account which follows in the text draws heavily on this source, and on Fox (1974, chs 5–9). Yip (1969, Part III) also is still useful on the restriction schemes as they affected Malaya, and Gill Burke's (1990) lively essay on tin restriction is well worth reading. See also Knorr (1945).

10. This is according to Hillman (1990b, p. 300). Fox (1974, pp. 157 and 172), however, records unsuccessful negotiations with the Yunnan authorities prior to the second agreement, and an unsuccessful approach to China prior to the third.

11. See Hillman (1988a, p. 91, and 1988b, p. 243) for further comment, and for supporting quotations from official records.

12. This is because inflation may be generated (in which case depreciation in the nominal exchange rate does not result in an equivalent fall in the real exchange rate) or because the cuts are difficult to sustain politically and socially. Devaluation, indeed, does not normally allow painless adjustment, especially if wages and living standards are already low. These are now fairly standard propositions from international economics. If devaluation is to be effective, real absorption must be reduced in relation to real output in order to reduce imports and free resources for export. For a succinct statement of the issues, see R. Dornbusch in Williamson (1983, pp. 223–30); for a fuller account, see Dornbusch and Helmers (1988, chs 1–6).

CHAPTER 4

The Second World War and the Early Post-war Years, 1945–60

The world tin industry had entered the second world war with the international cartel agreement still in force between the major exporters. These countries, except for Thailand and Bolivia, were still colonies. Britain as a colonial power exerted great influence over the industry, and the British Colonial Office had played a large role in administering the tin agreements. Whilst the early post-war period saw political independence for the former colonies, British control in both production and smelting remained important. The growing economic power of the United States, however, affected the tin market. The tin trade during the war was tightly controlled by the Allies, while the Japanese occupied the South-East Asian tinfields. The post-war international tin agreements did not continue the pre-war system. They were slow to start and weaker in content, thanks in large measure to US opposition. The US also exerted influence through its stockpiling policy, and through its willingness to take aggressive action when it felt its interests were threatened. In Indonesia political independence was followed by the nationalisation of the remaining Dutch interests in tin, and in Bolivia the three main tin companies were nationalised. In other producers change was more gradual, and an essentially pre-war structure remained intact for longer.

This chapter first looks at the wartime tin market, and then at the formation of the first post-war international tin agreement. It discusses nationalisation in Indonesia and Bolivia, the gradual changes in Malaysia and Thailand, and developments in Nigeria and other producers.

THE WARTIME TIN MARKET[1]

From the start of the war, tin came under formal control by the Allied governments. In the UK tin supplies were dealt with by the Ministry of Supply through its Nonferrous Metals Control Division. Tin prices were frozen by the Ministry of Supply, which became the sole UK buyer and seller of tin after shutting down the London Metal Exchange in December 1941 (Freund, 1981, p. 136). In the United States the government Metal Reserve Corporation became from 1940 the sole buyer of tin, establishing a price for the metal in the US market. Among the Axis powers, Germany was cut off from the tin supplies it had previously been purchasing in South-East Asia, and its occupation of Western Europe removed some smelting capacity (in particular the Arnhem smelter) from Allied use. Even after Japan had occupied the South-East Asian tinfields in 1942, shipping difficulties impeded tin supplies to Germany, while the South-East Asian capacity was far in excess of Japan's own needs.

The loss of South-East Asian supplies led to an intensification of UK and US control over what remained to the Allies of the tin industry. There were measures to control not only prices, but also usage and international allocation. American wartime measures were effective in requiring economy of tin use, pushing producers into introducing electrolyic tinplating and reducing the tin content in solder manufacture (Robertson, 1982, pp. 87, 97).

Much of the control apparatus remained in being into the late 1940s, after the war had ended. Even before the war the Americans had been considering building stocks of strategic minerals – tin was of particular concern because, in contrast to most other major metals, the US had neither supplies of its own, nor (except potentially in the case of Bolivia) easy access to sources from nearby countries. Some minor stockpiling had occured in the 1930s, and in 1940–1 the Americans started building a significant stockpile through direct contracts with the International Tin Committee. By 1942 a target of 100,000 tons had been largely achieved (Fox, 1974, p. 226–7).

With the loss of South-East Asia, which had produced over 140,000 tons of tin-in-concentrates in 1940, Bolivian tin assumed great importance. Bolivia raised its annual output for most of the war years to around 40,000 tons (see Table 4.1), a figure unmatched before or since except for 1928–9 (ITRDC, 1937). Nigeria also raised its wartime annual output to an all-time peak of some 12,000 tons, while the Belgian Congo increased to about 17,000 tons. Competition for Bolivian supplies between Britain and the United States reflected a desire on the part of the Americans, as the world's largest consumer of tin, to weaken the very tight hold the British exercised over the industry.[2] This hold was exercised not only through British interests in Malayan and Nigerian tin-mining (and Thai and Burmese tin too to some extent), and Britain's role as a colonial power, but through smelting. While British investors controlled about 30 per cent of world tin production, 70 per cent of the world's tin was smelted by smelters in the British

TABLE 4.1   Production of tin-in-concentrates, by country, 1941–60

(000 tons)	1941	1944	1946	1948	1952	1955	1960
Austria, Czechoslovakia and Germany	0.60	0.63	0.15	NA	0.4	0.7	0.7
Portugal, Spain, and Italy	3.1	1.1	1.4	1.0	2.4	2.3	0.9
USSR	NA	NA	NA	NA	NA	NA	NA
UK	1.5	1.3	0.79	1.3	0.9	1.1	1.2
Nigeria	12.1	12.5	10.3	9.2	8.5	8.4	7.8
South Africa	0.52	0.50	0.49	0.46	1.0	1.3	1.3
Belgian Congo/ Zaire	15.7	16.9	14.2	12.9	12.3	13.5	9.4
Brazil	–	0.15	0.27	0.2	–	–	1.6
Bolivia	42.1	38.7	37.6	37.3	33.0	28.9	19.7
Burma	5.6	0.5e	0.34	1.1	1.1	1.1	1.0
China	6.4	2.0	2.7	4.9	5.6	12.2	28.5
Indo-China	1.3	0.36	–	0.03	0.2	0.3	0.4
Japan	2.2	0.37	0.08	0.12	0.7	0.9	0.9
Malaya/ Malaysia	79.4	9.3	8.4	44.8	58.9	63.2	52.8
Thailand	15.8	3.3	1.1	4.2	9.7	11.4	12.3
Dutch East Indies/ Indonesia	53.4	7.0	6.4	30.6	36.2	34.5	23.0
Australia	3.5	2.6	2.1	1.9	1.6	2.0	2.2
World	246.0	100.0	89.0	152.0	176.6	186.4	170.1

Sources and Notes
1. 1941–48 from ITSG (1949), except Brazil 1948 from Schmitz (1979).
2. 1960 figures for major producers are supplied by Malaysian Ministry of Primary Industries, from ITC sources. All other figures are from Schmitz (1979), converted from tonnes to tons.
3. 1952–60 figures for 'Austria, Czechoslovakia and Germany' are for Germany alone; 1952–60 'Indo-China' figures are for Laos; 1952–60 figures for 'Portugal, Spain and Italy' are for Portugal and Spain alone.

Empire, and these supplied over 80 per cent of the American market (Hillman, 1990a, p. 291). Some three quarters of Bolivian concentrates were smelted in the UK, mainly in the Williams, Harvey smelter owned by Patino's Consolidated Tin Smelters. For technical reasons, the (mainly high-grade) concentrates from Bolivia's hard-rock tin deposits were smelted best with the addition of alluvial concentrates. By means of the prohibitive export duties on concentrates from Malaya and Nigeria, Britain had denied supplies of alluvial concentrates to the United States and any other potential competitor in the smelting industry,[3] and insufficient high-grade concentrates from Bolivia's hard-rock deposits were available for another smelter to operate at anything approaching minimum efficient size (Hillman, 1990a, pp. 291–2). Britain's power in the industry had been exercised strongly through the inter-war tin cartel run by the International Tin Committee, and not, the Americans may have felt, to the interests of the United States.

During the war the United States opened a large smelter in Texas City which was operated by the Billiton company on behalf of the US government's Reconstruction Finance Corporation (of which the Metals Resources Corporation was a subsidiary). The Texas City smelter's size was gradually expanded to 90,000 tons, creating gross excess capacity in relation to the available supplies of concentrates (Fox, 1974, p. 202; Hillman, 1990a, pp. 297 and 309). Competition between the US and the UK led to protracted negotiations with Patino about prices (which in turn were greatly influenced by the effects of exchange control, which created differentials between tin sold for US dollars and tin sold for sterling). In the event, the British hold over Bolivian supplies was not broken during the war, and the Texas City smelter proved difficult to operate efficiently. However, the United States' intention to countervail foreign influence over American tin supplies would be realised later through its buying power on the international tin market, and by its stockpile policy in particular.

The structure of wartime control was continued to 1949 under the overall control of the Combined War Materials Board. The Ministry of Supply in Britain removed price control and ended its monopoly purchasing in November of that year (Freund, 1981, p. 205). However, as there seemed likely to be an increased demand for tin as Europe reconstructed, the Board's activities on allocation were prolonged and widened to a larger number of users by the establishment of a Combined Tin Committee to distribute tin metal between members.

THE ESTABLISHMENT OF THE POST-WAR TIN AGREEMENTS[4]

When the war ended there existed in principle the machinery to restart a tin agreement. The fourth of the international tin cartel agreements had been accepted by most of the major participants, and was due to come into force in 1942, when Malaya and the Dutch East Indies were overrun by the Japanese. Some discussions were held after the war in the International Tin Committee about the creation of a fifth agreement to run from 1947, but the idea collapsed when the British government made clear it would not sign any proposal for a new agreement. The ITC met for the last time in December 1946. The British apparently were well aware of American hostility, and instead proposed an international tin conference, which took place in October 1946. Out of this came an international tin study group (ITSG), which would act as a forum in which future difficulties confronting the industry could be discussed (Fox, 1974, pp. 195–8). The ITSG was paralleled by study groups in other commodities, including rubber, lead and zinc, cocoa and sugar.

Thinking on international commodity agreements, though dating back to work in the League of Nations in the 1930s, was more immediately influenced by the Havana Charter, signed after the conference in Havana in 1947–8 which was convened to discuss the setting up of an International Trade Organisation (ITO). The ITO would have paralleled the two other 'Bretton Woods' institutions, the World Bank and the International Monetary Fund, which were established after the 1944 Bretton Woods conference on the post-war world economic order. In fact, the ITO was never ratified, and eventually was replaced by a body more limited in its scope, the General Agreement on Tariffs and Trade (GATT). The Havana Charter nevertheless was an important background to the planning which went on inside the ITSG towards a new, post-war international agreement for tin.

Although early drafts of the agreements included the principle of allocating supplies between member countries, as in the Combined Tin Committee, the form which eventually was agreed was based on intervention by a buffer stock, which could be backed by export control if the buffer stock holdings of tin rose beyond a critical level (10,000 tons). This contrasts with the inter-war agreements, where buffer stock operations essentially had supplemented export control.[5] The agreement followed the Havana Charter's recognition of the possibility of a 'burdensome surplus' emerging, which might cause hardship and unemployment in producing countries. This was because the price elasticity of demand and supply for commodities was thought to be too low to bring about ready adjustment by market forces; thus production restriction would be justified. The agreement also had the aim of achieving price stability, and ensuring adequate tin supplies at reasonable prices (Fox, 1974, pp. 206–7, 246–7). The ITA set floor and ceiling prices at which the buffer stock manager had to intervene, and upper and lower ranges where he[6] could intervene. There was also a middle range where it was intended there would be no intervention, although this was soon reversed when the advantages of such increased discretion became clear. The first ITA (ITA1) had a buffer stock of 25,000 tons, which could be contributed in cash or in metal at the ITC's discretion; in negotiations after the war the Americans had called for a larger one. The second, third and fourth agreements would have buffer stocks of 20,000 tons, and only the fifth (in deference to the Americans as new members) would have a larger buffer stock.

The first ITA was agreed at the end of 1953, and ran from July 1956 to June 1961. Although the United States had been an important influence on the discussions, it did not join the first agreement (or indeed any other, except the fifth – see Chapters 5 and 6). Membership of ITA1 covered 90 per cent of world tin mine production outside of the USSR and China. Initially it covered only 40 per cent of (non-communist) world consumption, the US and West Germany not being members. West Germany eventually joined the fourth agreement in 1971, as also did the USSR.

The delay in agreeing on, and starting, the international tin agreement is explainable in part by the peculiarities of the post-war period, which saw American stockpiling of tin and the outbreak of the Korean war, and also the maintenance of some direct control, particularly in the USA. Other primary commodity groups, however, were even slower, and generally less successful in starting commodity agreements.

The possibility of a early post-war tin surplus was widely foreseen, in the sense that trends in consumption were expected to lag behind consumption once post-war rehabilitation of the industry was well underway. The short-run problem was masked until the mid 1950s, however, by the United States' stockpiling of tin. Stockpiling, already significant during the war, was given impetus by the growth of communism in China, culminating in the 1949 revolution, and by communist activity in Malaya and Vietnam such that the Americans developed a fear of losing their Asian supplies. General policy had been established in the Strategic and Critical Minerals Stock Piling Act of 1946 (Public Law 520), and in 1948 the Americans had indicated their desire for stockpiled materials to be sufficient for wartime peak consumption for a five-year supply interruption (Fox, 1974, p. 231).

The easing of tin control by the US and the UK from 1949, as rehabilitation had got underway in South-East Asia, was quickly followed by panic buying as the Korean war broke out, not least by US agencies. Rapid rises in price from mid-1950 to early 1951 were met with bitter American complaints about 'price gouging' by rapacious 'allies', although the Malayans argued they were producing as much as possible. The price rises were countered by a buying strike by the United States in 1951–2, when it returned to wartime restrictions on tin purchase and use, and ran down existing stocks. By early 1952 strategic buying of tin for the US stockpile had resumed, and by late 1952 private importing of tin into the US was again possible.

By 1954 the US stockpile was revealed by the Americans to have reached minimum acceptable levels, although some further increases were planned. Kept secret until 1961, its size was estimated by Fox (1974, p. 241) to have been 350,000 tons in 1957, some two year's world consumption.

US stockpiling having been virtually completed by the start of the first post-war international tin agreement, ITA1 was started in 1956 in the context of a substantial potential excess supply. In 1957, the market was hit by large exports of tin from the USSR, which peaked at 21,100 tons in 1958, a figure roughly equivalent to the USSR's imports of tin from China. The buffer stock quickly accumulated the export control triggering figure of 10,000 tons, and export control started in December 1957. Export control remained in force until the end of September 1960, during which time an agreement was reached with the Soviets to curtail their disruption. Control was tight, equivalent to only half of normal production in late 1958 (Fox, 1974, p. 293), and was generally well-enforced, except for some smuggling from Thailand. The fall in price in the early years of the agreement was a severe test for the ITA, and additional buffer stock contributions, as laid down in the agreement, were required from members. The buffer stock's resources had to be supplemented by special authorisation from the International Tin Council for the buffer stock manager to borrow from commercial banks funds to finance another 10,000 tons. This loan was repaid at a profit. Export control was maintained longer than strictly needed for immediate price stabilisation in order that the large stocks of tin in the buffer stock, including that financed by the special borrowing, could be liquidated for cash. Fox (1974, p. 280) argues, in retrospect, that control was maintained too long, with the unfortunate consequence for the 1960s that the buffer stock was denuded of tin from June 1961 onwards. Nevertheless, the international tin agreement had survived severe testing.

THE INDONESIAN AND BOLIVIAN NATIONALISATIONS

In 1952 the three largest mining groups in Bolivia – Patino, Hochschild and Aramayo – were nationalised by the new Nationalist Revolutionary Movement (MNR) government, a regime which had seized power with the support of city dwellers and militant, armed mineworkers. A national mining corporation, Comibol (Corporacion Minera de Bolivia), was formed to work these mines, which then accounted for about three-quarters of the country's tin output. In Indonesia after the war the Dutch company Billiton had been given a five-year contract to run the Bangka and Belitung island mines. The contract for Bangka was not renewed in 1953, and in 1958 the rights of Billiton on Belitung and Singkep islands were taken over by the government (Fox, 1974, p. 35).

Bolivia

The Bolivian nationalisation ran into severe problems. Comibol's tin production fell steadily from over 27,000 tonnes in 1952 to just over 15,000 in 1960. Its workforce grew from 24,000 in 1951 to 36,558 in 1956 and by the end of the 1950s labour productivity had fallen by about 60 per cent.[7] The rise in the workforce undoubtedly represented the power of the mineworkers union and a system of workers' control, where workers were able to take an active part in controlling management decisions. Some 200 foreign technicians are estimated to have left Bolivia after nationalisation (Zondag 1966, p. 88). In addition, a five-month delay between the announcement of the nationalisation and its implementation gave the former mine-owners an opportunity to repatriate capital and to work existing deposits to the maximum. The difficulty of smelting Bolivia's complex ores meant that concentrates continued to be sent to the Williams, Harvey smelter in the UK, owned by Consolidated Tin Smelters in which the largest Bolivian mineowner, Patino, had a substantial interest. This put Patino in a stronger position to claim compensation for the nationalisation (Ayub and Hashimoto, 1985, p. 15). From 1953–61 the Williams, Harvey smelter, owned by the Patino organisation, is said to have deducted nearly $20 million (in the form of a 10 per cent discount) as an 'advance' on indemnification payments for the nationalisation (Widyono, 1977, p. 24).

The fall in output was only partly a reflection of maladministration, however. It also reflected an almost total lack of reinvestment in the previous twenty years[8] and a failure to discover new deposits during that period, with the result that ore grades declined. Revenue from Comibol was creamed off, principally by a system of multiple exchange rates whereby the corporation had to sell its export proceeds to the authorities at an artificially low price. This was in order to finance development elsewhere, especially in the petroleum sector and in agriculture. The decline in Comibol's production must be set against a background of wider dislocation, caused partly by the mineral sector's own decline and partly as the short-term consequence of land reform. In the five years after tin nationalisation, GDP in Bolivia fell 20 per cent per capita and by 1956 the domestic price level had risen by over 2000 per cent compared with 1951.

An economic stabilisation policy was introduced in 1956, involving currency reform and a variety of decontrolling measures (including freeing the exchange rate) to achieve its objectives. Comibol was freed from the implicit taxation of the multiple exchange rates[9] and an export royalty replaced the previous income and export taxes. Simultaneously the private medium-sized miners were relieved of a similar obligation to sell their ore to the Mining Bank, an institution set up in 1936, and nationalised in 1939, to aid smaller miners *vis-à-vis* the 'big three' and large ore buyers (Geddes, 1972, p. 347).

The potentially beneficial effects of the 1956 stabilisation programme for Comibol were offset by the continuing low world price of tin, which, as already shown, had experienced a sharp decline in the mid-1950s as Soviet tin exports depressed the world markets and the USA ceased stockpiling; and by the ITC export control from the end of 1957 until late 1960. Overmanning and lack of investment – only $3 million in the whole of the 1950s – made for a continuing decline in output, which the large influx of foreign capital in 1961–3 under the so-called Triangular Operation for the Rehabilitation of the Nationalised Mines attempted to stem (see next chapter). During the late 1950s Comibol made losses of about $10 million per annum, even though from 1957 to 1963 it paid no taxes on its exports.[10]

Indonesia[11]

During the war the mining islands had been occupied by the Japanese from 1942 to 1945. Belitung, unlike Bangka, was handed back to the Dutch without any interregnum of Indonesian independent rule, although there were a series of labour disputes with Chinese mineworkers in the early period of rehabilitation of the mines. Under the Japanese occupation, when the mines on Bangka, Belitung and Singkep were run by the Mitsubishi company, production had been small. Shipping problems made the export of ore to Japan difficult, and much damage had been done to the mines by the Dutch prior to the Japanese occupation.

In 1948 a single management was formed to run all of the country's tin mines – the Netherlands Indies Tin Council, or Tin-raad. The Billiton Company's concession was renewed and the company was also asked to operate Bangka for five years on a profit-sharing basis. The government of newly-independent Indonesia in 1949 inherited the Dutch colonial government's five-eighths share in the Billiton Joint Mining Company. In 1953 the Indonesian government took over the tin industry on Bangka, and P. N. Tambang Timah Bangka was set up to run the island's mines. Unlike the plantations in Indonesia, the administration of the tin mines was not taken over by the military.

In 1958, when the Billiton company's concession on Belitung was terminated and the administration given over to P. N. Tambang Timah Belitung, Heidhues (1992) notes that there were almost as many Europeans working on Belitung as in the 1920s. Although they were given the option of staying on, almost all chose to be repatriated.

In the 1960s as part of General Suharto's 'New Order', the three separate state mining organisations of Bangka, Belitung and Singkep were combined into P. N. Tambang Timah, and later (in 1976) changed into a limited company P. T. Tambang Timah Persero.

The decline in Comibol's output after nationalisation in Bolivia is closely paralleled by the experience of the Indonesian takeover of the tin industry. The taking over of Billiton's remaining leases was part of a general expulsion of Dutch interests in 1957–8, and the outflow of Dutch personnel was significant, since they had been important in running the tin mines. Indonesia's tin exports would fall from about 23,000 tonnes in 1960 to under 13,000 in 1966, though this was in part due to the general dislocation of the economy under President Sukarno, with high inflation and an over-valued exchange rate.

MALAYSIA AND THAILAND

When the Japanese invaded northern Malaya in December 1941, and as Japanese troops pressed on down the peninsula towards Singapore, the retreating British forces carried out a scorched earth policy. As part of this some tin dredges were sunk, and mining equipment destroyed or damaged to prevent the industry falling intact into Japanese hands (Loh, 1988, pp. 57–8).

During the occupation, after a brief spell of direct administration by the Japanese authorities, the Western sector of the industry was put into the hands of various Japanese companies, including Mitsui which also operated the Straits Trading Company's tin smelter. Many Chinese-owned mines were also seized by the Japanese. By 1943 Malayan tin output was up to 26,000 tons, which was still very low compared to the 1940 figure of over 80,000 tons. By 1944 labour shortages had developed, as workers were shipped off for forced labour for the Japanese war effort and as others escaped to the jungle to avoid it. These shortages, combined with difficulties in obtaining mining supplies, led to output falling to some 9000 tons in 1944, and to 3000 in 1945, when there was also sabotage of mines by anti-Japanese partisans. By the end of the occupation, no dredges and only 45 Chinese mines were working (Yip, 1969, pp. 289–97).

A rehabilitation scheme was quickly organised for the industry by the British Ministry of Supply, aware of British needs for dollar earnings from tin exports, and loans were disbursed to miners to restart. Yip (1969, pp. 302–3) argues the scheme discriminated in favour of European mines by virtue of its stress on long-term prospects and the size of a mine's capital as criteria to determine the size of loan. Many mines used the loans to update their equipment (e.g. converting dredges from steam to electricity, and introducing earth-moving equipment on gravel pump mines). The declaration of an Emergency in 1948 in Malaya, in the face of a (mainly ethnic Chinese) communist insurrection, according to Yip may actually have improved an unsettled labour situation by removing activists in the labour unions. By 1949 when the rehabilitation scheme officially ended, Malayan output was back to nearly 55,000 tons, and the prospect of a world tin surplus was beginning to be feared.

In Thailand, lack of fuel and spares meant that most mining by dredges came to a halt during the war. Rapid rehabilitation after the war had brought 31 dredges back into operation by 1950, compared to a total of 45 in 1940. Of the 25 companies operating dredges in 1950, three were Thai companies[12] and the rest were British or Australian. Some local smelting was carried out in Thailand during the war, but had collapsed by 1950 (Ingram, 1971, pp. 100–1).

In 1950 some 64 per cent of Thai tin output was being produced by dredges, a similar proportion to the late 1930s (Ingram, 1971, p. 100). By the 1960s their share had fallen to under half (Fox, 1974, p. 41). In Malaysia, dredging's share of tin output stayed constant during the 1950s at almost exactly a half, while gravel pumping's remained at just over a third (Thoburn, 1977b, p. 87). Yip (1969, p. 365) suggests that little or no new foreign capital flowed into Malaysia for tin mining after the second world war. There was new dredge building in Malaysia in the post-war period however (Thoburn, 1973b), but financed within Malaysia. Within the dredging sector in Malaysia the pre-war structure of control continued, with Anglo–Oriental and other agencies acting as agents for many dredging companies. Boards of directors were mainly foreign (and predominantly British), and there was a system of interlocking directorships between companies. Nevertheless, underlying these changes, though not affecting control, was a substantial rise in local ownership of shares, mainly by Chinese Malaysians. According to research by Yip (1969, pp. 346–80) on the ownership of tin companies over the period 1954 to 1964, which spanned Malayan independence, institutional investors in the UK tended to sell their shares, which were purchased by local shareholders. During this period only one new public tin dredging company was set up – Selangor Dredging in 1963, the first such to be established by Malaysian Chinese interests. Several major dredging companies in Thailand continued to be associated with London Tin Corporation control. The tin dredging sectors of Malaysia and Thailand thus tended to retain much of their pre-war character, and their international links were to specialist tin companies, particularly London Tin/Anglo–Oriental. The world's major mining multinationals were still conspicuous by their absence.

NIGERIA

Nigeria during the war became important as a supplier of tin for the Allied war effort and as an earner of dollars for the UK to pay for American supplies of war material. With difficulties in supplying additional equipment to the mines, a forced labour policy was introduced by the colonial government in Nigeria in an attempt to increase tin output. This policy was associated with particularly harsh labour conditions, despite some minimum wage legislation introduced in 1942–3. Large increases in labour inputs, in the absence of available mining equipment, produced a substantially less than proportionate output increase. Great disruption was caused to the local peasant economy (Freund, 1981, ch. 5).

As in Malaysia, the twenty years or so after the war saw little change in the economic structure of the Nigerian tin industry. There was also little new investment other than some for replacement. Associated Tin Mines of Nigeria, controlled by the London Tin Corporation, continued to account for almost half of Nigerian output, and there was some consolidation among smaller Western companies. Nigerians were granted mining rights after the war, and Freund (1981, p. 217) records that by 1959 there were 11 Nigerian tin mine operators. Unions had started in the tinfields during the war, with the approval of the colonial government, and real wage gains were secured at least up to the 1950s. Later, especially, the unions would become enmeshed in the inter-ethnic conflicts which culminated in the Nigerian civil war in the 1960s. The colonial state showed a greater interest after the war in using the tin industry as a source of revenue, and the share of export royalty in total tin sales rose from its 1920s and 1930s level, of well under 10 per cent, to 10–20 per cent for much of the 1950s. Domestic smelting did not start in Nigeria until the 1960s (Freund, 1981, ch. 8).

OTHER PRODUCERS

The Belgian Congo (Zaire), having become an important producer during the war, maintained high output during the 1940s and into the 1950s (the last decade of its colonial status). Thereafter the output decline would be rapid.

In China in the early post-war period some 80–90 per cent of its tin continued to be mined in Yunnan province. The largest mines in that area were operated by the Yunnan Tin Corporation, which was taken over by the new communist government after 1949. A smelter had been established in 1935 by the Yunnan Tin Corporation (ITSG, 1949; ITC, 1960).[13]

In Burma the tin industry suffered extensive damage during the wartime Japanese occupation and did not recover. In 1960 the ITC reported that post-war rehabilitation still had made little progress. The Burmese government had a policy of reducing foreign involvement. By 1960 most tin was still coming from the group of larger mines which previously had been Western-owned (ITSG, 1949; ITC, 1960).

During the early post-war period, British and South African production remained low, and Australian production had yet to make the increases which would later give it a larger output than Nigeria and Zaire combined.

Brazil during the late 1940s and the 1950s remained a very minor producer of tin. International Tin Council publications date the country's production of tin 'on a commercial scale' to the discovery of tin in Minas Gerais state in 1943. The discovery of tin in Rondonia in the Amazon, though dated to 1950, did not result in production from that region until 1959, and initially then only in negligible quantities (Engel and Allen, 1979, pp. 63–4). Nevertheless, some developments took place which would be significant for the industry's later growth. As part of its industrialisation drive, the Brazilian government in the 1950s gave incentives for local production of tin based on imported ore, placing high tariffs against imports of tin metal and restricting the export of concentrates. Domestic tinplate production had started in 1948 at the steel town of Volta Redonda some 250 km from Sao Paulo, initially using hot-dipped methods, and electrolytic tinplating was started in 1955 (ITC, 1960, p. 94). A tin smelter had been set up by Cesbra, one of the companies which pioneered the growth of the tin industry, and opened in 1954 at Volta Redonda. Brazil remained a net importer of tin throughout this period. Growth of domestic smelting and tinplating depended on imports of tin concentrates from a variety of sources including Thailand at one stage, and some European countries.

OVERVIEW: 1940-60

As in the inter-war period, there was little change in the geographical structure of production during these years, although Bolivia and Nigeria temporarily grew in importance during the second world war when South-East Asia was overrun by the Japanese. The output expansion in Nigeria was achieved by the colonial government by means of a harsh policy of forced labour.

The tin industry was nationalised in independent Indonesia as part of a general expulsion of Dutch interests from the economy, although the state already was playing a large role in tin. In Bolivia, the three largest tin mining companies were nationalised, to form the state company Comibol. The Patino organisation, however, retained its extensive international interests in smelting. Both countries' tin industries suffered a sustained period of output falls, though Indonesia's would recover later through a programme of new investment.

During the war tin supplies were tightly controlled by the Allies, and the international tin cartel agreement was still notionally in force. The UK during the war strenuously defended its interest in smelting ore from Bolivia against American opposition. During the war too, the United States started stockpiling strategic materials, which later, in the case of tin, would give it an influence on the tin market which it lacked in the inter-war period.

The international tin agreements of the inter-war period were allowed to lapse in the face of American opposition to them. The post-war tin agreements did not start until 1956, though even this was quicker than for other major commodities. As in the inter-war period, there was both provision for export control and a buffer stock. There were great fluctuations in price, including the Korean war boom, and the ITA was kept active. Erratic export sales by the USSR at times exerted a disruptive influence on the market.

In Malaysia and Thailand, the economic structure of the tin-mining industry appeared to remain intact, with output divided between a foreign-owned dredging sector and a locally-owned gravel pumping sector. However, there was significant growth in the local ownership of shares in foreign companies, though this was not accompanied by control. The London Tin Corporation remained an important player, and there was still minimal involvement in mining and smelting by tin-using companies.

NOTES

1. These details of wartime control come mainly from Fox (1974, ch. 9) and Hillman (1990a).
2. This point is strongly argued by Hillman (1990a).
3. However, as a goodwill gesture, the export duty on exports of Nigerian tin concentrates to the United States was lifted at the start of the second world war (Freund, 1981, p. 136). Nigeria at that time, unlike Malaya, did not have its own smelting capacity.
4. Fox (1974) is the main source for this section. Redzwan (1985) is also useful.
5. However, Fox (1974, p. 269) observes that in the first three post-war agreements, the buffer stock was the lesser weapon in the ITC's armoury – presumably because the buffer stock could quickly accumulate enough tin to justify export control in the event of marked excess supply.
6. There was never a female buffer stock manager.
7. These details are from Zondag (1966). Geddes (1972) is useful on the background to nationalisation, as also are Ayub and Hashimoto (1985).
8. This point is made by Geddes (1972, p. 258), Zondag (1966, p. 87), Gillis (1978, p. 28) and Ayub and Hashimoto (1985, pp. 13–14).
9. Note though that this system was not a creation of the MNR government. It had been used to tax the mining sector since the 1930s (Geddes, 1972, ch. 22).
10. However, according to the estimates of a former general manager of Comibol, the corporation made an average annual profit of about $10 million (on a turnover of under $100 million) during the early years after nationalisation (Zondag, 1965, pp. 84–5).
11. This section draws mainly on Heidhues (1991 and 1992).
12. Ingram (1971) unfortunately gives no further details, so it is not clear whether these are indigenous Thai dredging companies or merely foreign companies registered in Thailand. Given that there was no local publicly-owned dredging company set up in Malaysia until the 1960s, these Thai companies were probably foreign. The 1960 ITC *Statistical Yearbook* says that all dredges in Thailand were foreign-owned.
13. ITSG (1950) comments that virtually nothing has been known about the Chinese tin mines since the communist takeover. In Jones' (1925, pp. 240–5) account of the world tin industry, the mines in Yunnan were described as alluvial, worked by simple surface methods. By the time Fox (1974, p. 82) was writing they were described as underground, as they are today.

CHAPTER 5

The 1960s and 1970s
High Prices and Bargaining over Mineral Rents

The 1960s, and more especially the 1970s, were in some respects the high point of the tin industry's history, at least from the producers' viewpoint. Real tin prices, which had been constant in their long-term trend for virtually the whole of the century, rose gradually in the 1960s and more substantially in the 1970s (see Figure 1.1), although the price rise from the producers' viewpoint was not as large as the price rise faced by consumers (see Appendix). Even by the end of the 1970s, few commentators anticipated the collapse which was soon to occur. Major mining multinationals started to enter the industry, previously the preserve of specialist tin companies, as also did some oil companies, mirroring events in other metals. At the same time, even market-orientated countries such as Malaysia and Thailand started to become more nationalistic towards foreign investors, encouraged not only by the prevailing interest among less developed countries in a New International Economic Order for commodity producers but also by the very high mineral rents to be bargained for. This chapter starts with an account of the tin market, including the operations of the ITC and the US strategic stockpile disposals. It then looks at changes in industry structure with special regard to the participation of multinational companies, at the degree of forward and backward integration in the industry, and also at the growth of tin mining in Brazil. The third section is on changes in producer policy, especially during the 1970s; it considers first Malaysia and Thailand, then the operation of state enterprises in Bolivia and Indonesia, and some minor producers. While the first three sections broadly continue the format and coverage of earlier chapters, the final section changes focus: it moves from a chronological to a functional approach and presents a study of the cumulative impact of tin exports on economic development in the main producing countries in the 1970s.[1] The

1970s is an appropriate period for such a study since these development effects were then at their greatest, and data were available to allow the influence of policy to be observed. Thereafter, much of the story will be one of decline.

THE TIN MARKET

A summary of activities in the tin market is given in Table 5.1, which is extended up to 1984 in order to show the run up to the 1985 tin collapse.[2] Figures for 1985 and beyond are given in Chapter 6.

The tin surplus of the late 1950s had been eliminated by the start of the second International Tin Agreement, (ITA2) (1961–6), though export control remained in force until the end of September 1960. The assumption at the start of ITA2 was that of shortage (Fox, 1974, Ch. 15).

At this time the US stockpile of tin, having been built up to 348,000 tons, was seen by the American government to be in excess of long-term strategic needs, and the disposal of a surplus of 164,000 tons (later 148,000) was decided in principle. William Fox records protracted discussions between the ITC and the American authorities about the disposal of the surplus, with agreement being reached in October 1966 that the US would moderate its tin sales if these were inconsistent with the ITC's operations. In fact, initial disposals were made at a time of increasing shortage. By late 1963 and through 1964, the GSA made much larger sales than originally planned for the period, and the ITC 'remained a passive witness to events and prices which it could not now influence'. (Fox, 1974, p. 340). Over the 1963–6 period the GSA disposed of about half of its surplus. At the peak of stockpile disposals in 1964, the GSA's sales were about as large as the combined production of Indonesia and Thailand. Fox concludes that GSA sales over this period closely matched the statistical gap between expected production and consumption. In this sense the stockpile disposals served to defend the ITA ceiling price, though the Americans had resisted this as an explicit policy. By 1967 the disposal programme was being wound down, and large disposals were only made again during the 1973–4 boom.

The ITC was active in the market 1967–73, and imposed export controls from September 1968 to December 1969. It bought large quantities of tin in 1975 after the boom, backed with export control in much of 1975–6. Rather surprisingly, there was also export control from mid-January to September 1973 (Redzwan, 1985). The real price of tin, though it fell sharply in 1975, remained significantly higher than in the 1960s. By late 1976 the amount of tin the GSA was authorised to sell was nearing exhaustion. Chinese sales, which had revived in the early 1970s (though not to their early 1960s peak) had also tailed off, and ITA stocks were low

TABLE 5.1   World supply and demand for tin, 1960–84

Supply (tonnes)

Tonnes	Production	GSA sales	Chinese sales	ITC (purchases) sales	Total supply
1960	139,000	–	26,200	20	165,220
1961	139,000	–	21,000	10,191	170,191
1962	144,000	1,400	20,000	(3,322)	162,078
1963	144,000	10,626	17,300	3,322	175,248
1964	150,000	31,147	12,200	–	193,347
1965	154,000	21,733	10,100	–	185,833
1966	167,000	16,276	7,300	(36)	190,540
1967	174,000	6,146	5,800	(4,795)	181,151
1968	184,000	3,495	5,300	(6,640)	186,155
1969	179,000	2,048	5,900	6,807	193,755
1970	185,000	3,038	7,700	3,432	199,170
1971	187,000	1,736	10,800	(5,405)	194,131
1972	195,000	361	11,600	(5,842)	201,119
1973	188,000	19,949	11,700	11,478	231,127
1974	183,000	23,137	11,000	859	217,996
1975	181,000	575	15,200	(19,929)	176,846
1976	179,000	3,546	7,150	19,265	208,961
1977	189,000	2,635	2,388	806	194,779
1978	197,000	326	3,215	–	200,541
1979	200,000	–	3,291	–	203,291
1980	200,900	25	3.912	–	204,837
1981	204,700	5,920	4,722	(1,000)	214.342
1982	190,100	4,172	4,345	(51,700)	146,917
1983	159,000	2,865	3,170	(2,415)	162,620
1984	162,000	2,397	2,563	(6,727)	160,233

(Williamson, 1984). Prices rose sharply, leaving the ITC inactive and unable to defend the ceiling price for the rest of the 1970s. Non-communist world production increased to 200,000 tonnes by 1979. Of the main producers, as Table 5.2 shows, Thai production rose sharply, while that of Indonesia recovered from its post-nationalisation decline. In Malaysia, however, then the largest producer, production actually declined, mainly due to a large fall in output in the gravel pump sector. This decline was blamed by the industry on 'penal' rates of taxation, but probably it had more to do with difficulties in securing mining leases. Tin mining was extremely profitable in Malaysia at late 1970s prices, even at the then tax rates and with the high price of oil products (a major input) (Thoburn, 1978b).

World tin consumption fell after 1973 and real tin prices in the late

		Demand (tonnes)			
				Market	True
		USSR	Total	surplus	surplus
Tonnes	Consumption	imports	demand	(deficit)	(deficit)
1960	170,000	9,300	179,300	(14,080)	(14,100)
1961	166,000	9,200	175,200	(5,009)	(15,200)
1962	166,000	9,800	175,800	(13,722)	(10,400)
1963	169,000	7,800	176,800	(1,552)	(4,874)
1964	177,000	5,500	182,500	10,847	10,847
1965	173,000	5,800	178,800	7,033	7,033
1966	176,000	4,800	180,800	9,740	9,776
1967	175,000	5,700	180,700	451	5,246
1968	180,000	7,100	187,100	(945)	5,695
1969	187,000	6,800	193,800	(45)	(6,852)
1970	185,000	8,300	193,300	5,870	2,438
1971	189,000	4,400	193,400	731	6,136
1972	192,000	4,200	196,200	4,919	10,761
1973	213,000	4,000	217,000	14,127	2,649
1974	200,000	5,200	205,200	12,796	11,937
1975	174,000	9,700	183,700	(6,854)	13,075
1976	194,000	11,400	205,400	3,561	(15,704)
1977	185,000	11,700	196,700	(1,921)	(2,727)
1978	185,000	16,300	201,300	(759)	(759)
1979	186,000	16,300	202,300	991	991
1980	174,000	16,900	190,900	13,397	13,937
1981	163,000	18,500	181,500	32,842	33.842
1982	157,100	16,100	173,200	(26,283)	25,417
1983	155,100	12,483	167,583	(4,963)	(2,548)
1984	166,660	13,096	179,696	(19,646)	(12,736)

Sources and Notes
1. Figures from Williamson (1984), which are mainly from ITC sources.
2. Williamson's estimated figures for 1983 and 1984 have been updated from ITC
   (1986)

1970s continued the upward trend started in the 1960s. The price of aluminium, tin's major competitor, merely levelled off after having fallen for more than half a century (Humphreys, 1982, p. 217). The early 1980s recession, following the second oil price shock, reduced consumption sharply. The difficulties of the International Tin Agreement in the 1980s, and its eventual demise, are dealt with in Chapter 6.

There has been much academic interest in the effectiveness of the tin agreement in reducing price fluctuations. A widely-quoted study is that by Smith and Schink (1976), who estimated the extent to which the ITA and the US stockpile disposals influenced market prices and producer revenue for the period 1956 to 1973. Smith and Schink's simulations suggest that average annual changes in real price (prices deflated by the UN index of

TABLE 5.2   Production of tin-in-concentrates, by country, 1961–79

('000 tonnes)	1961	1964	1967	1970	1972	1976	1979
Germany	0.7	1.0	1.0	–	–	1.3	NA
Portugal and Spain	0.9	0.8	0.8	0.8	1.0	0.8	0.7
UK	1.2	1.2	1.5	1.7	3.3	3.3	2.7
USSR	NA	22.0	23.0	NA	NA	16.0	18.0
Nigeria	7.8	8.9	9.5	8.0	6.7	3.7	2.9
South Africa	1.5	1.6	1.8	2.0	2.1	2.4	2.7
Zaire	6.7	6.6	7.1	6.5	6.0	3.9	3.3
Bolivia	20.7	24.6	27.7	30.1	32.4	30.5	27.8
Brazil	0.6	1.2	1.8	3.7	2.9	5.4	6.6
Burma/Myanmar	1.0	0.6	0.3	0.3	0.5	0.8	1.2
China	30.0	25.0	20.0	NA	NA	22.0	17.0
Laos	0.4	0.3	0.5	0.6	0.8	0.6	0.6
Japan	0.9	0.8	1.2	0.8	0.9	0.6	0.7
Malaysia	56.0	60.0	72.1	73.8	76.8	63.4	63.0
Thailand	13.3	15.8	22.8	21.8	22.1	20.4	34.0
Indonesia	18.6	16.6	13.8	19.1	21.8	23.4	29.4
Australia	2.8	3.8	5.8	8.8	12.0	10.4	12.6
World	169.4	175.6	195.2	205.9	215.3	195.8	238.8

Sources and Notes
1. All 1979 figures are from UNCTAD (1992a). For other years: minor producers,
   China, USSR and world totals are from Schmitz (1979). Other figures are supplied by the Malaysian Ministry of Primary Industries, from ITC sources.

dollar price of exports by LDCs, adjusted for changes in the pound/dollar
exchange rate) in the absence of ITA and GSA activity would have been 10.4
per cent. The ITA reduced them only to 9.4 per cent (though the model
could not simulate ITA production control), whereas GSA activity reduced
them to 6.9 per cent. The GSA operations reduced (discounted) producer
income by about 5 per cent over the period, presumably by creaming off
the top of the market. The possibility (well known to the commodity control literature) that the ITA itself might have reduced producer income
(because price control in a demand shift market, assuming the supply function has a positive slope, reduces price when supply is high and raises it
only when supply is low) did not occur. Of course, as Gilbert (1987, p.
611) has observed, the Smith and Schink test is misconceived in so far as
the ITC left the main responsibility for preventing shortages to stockpile disposals. Fox (1974, p. 308) suggests that members of the ITC believed there
was a maximum price above which the Americans would not allow the
market to go, but he himself is sceptical about this view. Implicitly this
would have left much of the defence of the ceiling to the GSA, even
though the Americans did not accept this explicitly. As is shown above, this
relation worked well in the shortages of the early/mid-1960s, but in the late

1970s when the ITC had run out of tin, disposals from the stockpile were minimal. This was in spite of having a larger buffer stock under the fifth Agreement, and was partly caused by lack of build up of stock as a result of export control in 1975–6.

During the 1960s and 1970s there were some changes in the nature of the agreements. ITA3 (1966–71) added the UNCTAD goals of growth and increased export earnings of producing developing countries. ITA4 (1971–6) saw an increase in the relative importance of the buffer stock compared to export control (Fox, 1974, p. 270–2), and the buffer stock manager was allowed to sell above the ceiling price. ITA5 (1976–81/2) was the only tin agreement which the United States joined, under President Carter. It operated with a much larger buffer stock than previous Agreements (40,000 tonnes), partly, it is thought, to mollify American objections to export control.

An important innovation was the establishment in 1974 of the Economic and Price Review Panel. The EPRP's task was to study costs of production in member countries. Its work included the major study of costs and returns to investment made by Engel and Allen (1979). From 1975, costs became the main consideration in the determination of the ITA price range (Redzwan, 1985), and in the late 1970s and early 1980s 'costs' formed the basis of much dispute between producers and consumers, as discussed in Chapter 6. Over the period May 1976 to October 1981 the floor support price was progressively raised from M\$16.5/kg to 29.15/kg (Williamson, 1984).

According to estimates for the period 1956–77 by Hallwood (1979), the tin agreements were broadly self-financing.

CHANGES IN INDUSTRY STRUCTURE

Changes in Participation by Multinational Companies

In Malaysia and Thailand various changes occurred which affected the relative bargaining positions of the companies and the host-country prior to the major policy changes in the mid-1970s which are discussed later in the chapter. First, several new multinationals entered the industry. In Malaysia, Conzinc Rio Tinto Malaysia (CRM), an Australian-owned subsidiary of the Rio Tinto Zinc Corporation (RTZ), one of the world's largest mining companies, started with a single dredging operation in the early 1970s, and embarked subsequently on several joint ventures with individual state government mining organisations. More important, the South African-controlled mining house Charter Consolidated, having bought into Malaysian tin in 1965, by 1975 controlled companies producing nearly 10 per cent of total Malaysian tin output; thus rivalling the London Tin Corporation, whose companies accounted for 15 to 17 per cent.[3] Charter subsequently became the minority partner in the federal Malaysia Mining Corporation (of which more later in the chapter).

In Thailand, the Dutch company Billiton, which originated in colonial Indonesia, entered the tin industry in partnership originally with the American company Union Carbide (UC). UC was the second largest chemical company in the USA, and owned a variety of mining interests in the Third World. UC had been especially interested apparently in certain byproducts of tin mining and tin smelting. In 1965 UC had started the construction of the Thaisarco tin smelter in partnership with a Thai company which had previously secured a large offshore mining concession. Billiton was invited to join the venture for its mining expertise in about 1970, the Thai partner having left. Mining and exploration operations were organised under two companies (TEMCO and EMCO, about which more later) with substantial offshore leases. UC left in 1974, following difficulties with (and eventually revocation of) the mining leases, giving Billiton 100 per cent ownership of the mining operations and the smelter.

The entry of RTZ (CRM) and Charter, and Billiton too to some extent, represented a significant change in industry structure and in the foreign bargaining position. In the early twentieth century it was quite common for mining companies to be centred on a single country, but the subsequent pattern has been of diversification using retained profits.[4] While the LDC resulted from the horizontal integration of tin-mining companies in several countries, even by the early 1970s it had not diversified significantly beyond tin. RTZ in contrast, the largest mining company in Britain, had interests in almost every metal in a wide variety of LDCs and industrial countries, and substantial interests in smelting in various non-ferrous metals. Its interest in tin was fairly new, however. RTZ's stake in Malaysia through CRM remained small, and it is Charter's which is of greater interest. Charter Consolidated was formed in 1965 out of several disparate companies in the Anglo–American Corporation group, the South African mining house.[5] Anglo held in 1979 a controlling 35.6 per cent of the equity of Charter (Charter Consolidated *AR*, 1979), which was one of Anglo's main vehicles for diversification beyond South Africa. Like RTZ, Charter had interests in a wide range of minerals, including lead, zinc, wolfram, potash and gold. Its activities in South-East Asia generated only 9 per cent of its 1979 revenue.

Billiton, too, was a much more diversified company than the LDC . During the 1930s it spread from Indonesian tin into investment in bauxite. After its exclusion from Indonesia in 1958 it moved into lead and zinc. In 1970 it was taken over by Royal Dutch Shell as part of Shell's diversification into minerals, a move paralleled by other oil companies (UNIDO, 1979, p. 203).

Billiton's ownership of the Thaisarco smelter in Thailand supported a major marketing operation by Billiton which handled not only almost all of Thailand's production, but also about 18,000 tonnes from Malaysia, Bolivia and China; all in all about 20 per cent of world tin trade.

In contrast to these new and diversified multinationals, the London Tin Corporation in 1974 still had 85 per cent of its investment in tin, of which Malaysia accounted for 75 percentage points.[6] Other metals and oil accounted for another 10 percentage points (Zorn and Leigh-Hunt, 1974).

The second type of change in Thailand and Malaysia concerns the inter-action between foreign companies and the local sectors. In both countries the range of techniques in use remained similar to that between the two world wars, apart from the growth of small boat mining in Thailand. In both countries the industry remained fragmented. In Malaysia gravel pump operations comprised about 90 per cent of the country's total of 800 tin mines, and small-scale operators in Thailand were over 90 per cent of the total.[7] In both countries, dredging, the main area of foreign involvement,[8] declined in relative importance, falling in Malaysia from about 55 per cent at the end of the 1950s to just over 30 per cent in 1975, and in Thailand from 65 per cent in 1950 to 16.6 per cent in 1976. In Malaysia the dredging decline was matched by a rise in gravel pumping's share, and in Thailand by a rise in gravel pumping and by the rapid growth of small boat mining from the mid 1970s. The relative profitabilities of dredging and gravel pumping during the post-war period would be difficult to establish without substantial historical research. It has been shown that gravel pumping's profitability is more price sensitive than that of dredging, and their relative rates of return on investment switch within realistic price ranges.[9] However, as reference back to Figure 1.1 will show, there does not seem to have been a clearly-defined upward trend in the real tin price until the early 1970s to explain gravel pumping's relative expansion; and one would expect rising real wages in Malaysia and Thailand to have favoured the capital-intensive technique. To some extent it may be an indication that new deposits were small and therefore unsuitable for dredging, or simply that there had been little exploration for new deposits. The latter factor was greatly stressed in Malaysian discussions in the 1970s. The failure of Malaysian output to expand, despite the high real price of the 1970s, was attributed to the adverse effect on investment incentives of 'penal' rates of taxation (Thoburn, 1978b). As a result of the lack of exploration, much gravel pumping was on land previously worked by dredges.[10]

In fact, much new dredging investment had taken place in Malaysia since the second world war and, until the output decline from 1972, it had been in the context of a rising output trend.[11] In Thailand much less dredging investment took place after the second world war and Thailand largely missed out on the really large capacity 'second generation' of dredges built in Malaysia.[12] A striking change in Thailand was the entry of local firms into dredging using secondhand equipment. Of the seventeen dredges in operation in 1976, twelve were working inland, and of the latter, nine were in the control of local firms (probably wholly Thai-owned). Some estimates based on company reports suggest that in 1977 only about 40 per cent of inland dredging output was produced by companies not wholly local, though until 1978 the entire offshore dredging output was produced by foreign companies.[13]

In Thailand, besides the Billiton–Union Carbide operation, there remained by the early 1970s only two large foreign groups, both of long standing. The 'TASK' group, comprising Tongkah Harbour, Aokam, Southern Kinta and Kamunting was under the control of London Tin Corporation; Southern Kinta and Kamunting also operated in Malaysia.[14] The other group was the Siamese Tin Syndicate and Bangrin who were grouped under the Fairmont State company in Thailand,[15] which had some Cornish links.[16]

In Malaysia direct local entry into dredging by the private sector was much less frequent. Indeed, there are only two known cases. In the first, Selangor Dredging, a public company, was floated by Malaysian Chinese in the 1960s; in the other a large gravel pumping company bought secondhand dredges and was operating two properties in the 1960s and 1970s in Selangor and Perak. Thus in Malaysia prior to the takeover of London Tin Corporation by the federal government (and the formation of the Malaysia Mining Corporation) the dredging sector was still quite concentrated. The LTC owned over 25 per cent of the shares of seven dredging companies and over 10 per cent in a further three (Thoburn, 1977b, p. 127). Although in no case was majority ownership held, control could effectively be exercised since other ownership was dispersed, and the instruments of control were in providing management services on the pattern established in pre-war years through Anglo–Oriental. The high degree of concentration within dredging in Malaysia is further illustrated by the fact that twelve companies, producing nearly 80 per cent of the country's dredging output, had as secretaries or managers either Anglo–Oriental, Osborne and Chappel or Charter Consolidated (or its subsidiary Associated Mines) personnel (Thoburn, 1977b p. 127). Lim Mah Hui's (1981, pp. 98–102) study of ownership and control in the Malaysian economy in the mid-1970s regarded the London Tin Corporation–Charter Consolidated grouping as being extremely tightly knit. He also saw it as highly homogeneous, with the interlocking directorships within its core being entirely between tin-mining companies, except for the Chartered Bank. In spite of its British origins, the multinationality of Osborne and Chappel is limited in the sense that its activities appeared to be mainly in Malaysia. However, it had had some links with LTC through interlocking directorships, and branched out into mining consultancy overseas.

In addition to the post-war changes in control, there were some changes in ownership. Increases in local purchases of shares in locally incorporated

The 1960s and 1970s

tin dredging companies in Malaysia is well documented over the period 1954–64, as noted in Chapter 4, and by 1970 over a third of all dredging company shares in Malaysia were locally-owned (Thoburn, 1977b, p. 89).

In Indonesia, foreign investment in tin was allowed after the change of regime in the 1960s, under the 1967 Foreign Capital Investment Law. Only three multinationals entered the industry.[17, 18] Billiton re-entered Indonesia in 1968 with a forty-year contract to prospect and mine offshore in the Riau archipelago near Singkep. In 1979 it started mining with the world's largest tin dredge. Billiton was (and still is) wholly owned by Royal Dutch Shell, though its Indonesian subsidiary was 10 per cent owned by Timah, and it had to offer 25 per cent of its equity to Indonesians over ten years. Broken Hill Proprietary (BHP), the second foreign company, whose contract dated from 1971, re-opened an old lode, the only underground mine in the Indonesian tin industry, on Belitung island. Broken Hill's parent company, of the same name, was one of the three largest mining companies in Australia (McKern, 1976, p. 13), and also a major manufacturer of steel and tinplate (TI, January 1980). The Indonesian operation seems to have yielded little profit and Timah did not exercise its option to take up an equity stake. Whereas BHP and Billiton offered specialised mining expertise in areas where Timah to some extent still lacked it, the third venture, Koba Tin, operated a mainly gravel pump operation. It was apparently extremely profitable and Timah held 25 per cent of its equity with the option of purchasing up to 45 per cent at 2 per cent per annum. The foreign owners were the Australian companies Colonial Sugar Refining and Blue Metal Industries, the former an active mining company. All three multinationals in Indonesia operated under 'second generation' contracts, basically similar to those in other mining.[19] Tin was said to be closed to foreign investment by the late 1970s, and in any case any new company would have been subject to the more stringent conditions of a 'third generation' of contracts.[20]

Among the minor producers, at the beginning of the 1970s Nigerian tin production was still dominated by foreign companies. Amalgamated Tin Mines of Nigeria, in which the London Tin Corporation had a 30 per cent controlling interest, produced about 45 per cent of the country's output, and identifiably British companies in total produced about two-thirds (Zorn and Leigh-Hunt, 1974). In Zaire, by the early 1970s (according to Fox, 1974, p. 53) there were only four groups of importance, all working alluvial concessions and all Belgian-owned. In 1976 three groups joined to form Sominki, producing about two-thirds of current output. Geomines, the other group, had become Zairetain, producing about one-third of the country's output, with the Zaire government taking part of the equity. Geomines' Belgian ownership was apparently independent, whereas Sominki was part of the Cofimines group. Cofimines was in turn part of the

Baron Empain Schneider group that had diversified into steel, transport and power and employed over 100,000 people.[21]

In Australia there was a large-scale revival of the industry in the 1960s as part of a more general minerals boom. In 1965 Australia became a net exporter of tin for the first time since the end of the second world war. This was associated with the activities of the multinationals Consolidated Goldfields of the UK and Cominco of Canada (Fox, 1974, pp. 73–5). CGF, like Anglo–American, was of South African origin and tin accounted for 18 per cent of its profits in according to its annual report. CGF's Renison underground mine in Australia produced over 5000 tons a year, the largest output of any private tin mine in the world, and almost half of Australia's production. CGF also had substantial minority ownership in the Rooiberg underground mine in South Africa,[22] the country's largest, and was instrumental in opening the Wheal Jane tin mine in the UK in the 1970s. Cominco owed the Aberfoyle mine, Australia's second largest. The Anglo–American Corporation, which owned Charter Consolidated, acquired early in 1980 a 25 per cent interest in CGF, though claiming it did not wish to exercise control. British Petroleum also took a stake in the Anglo group by its bid in mid 1980 for Selection Trust, in which Charter had 27 per cent (*The Observer*, London, 13 July 1980).

Myanmar (Burma), which had nationalised mineral exploitation including tin in the 1960s, placed all tin mining under the control of 'Mining Corporation Two', set up in 1972, which produced nearly 90 per cent of the country's tin by gravel pumping, with tributors producing the rest by panning (*TI*, October 1986).

Backward and Forward Integration

Integration backwards by multinational tin mining companies into mining supplies and mining equipment was very rare. Anglo–Oriental's former dredge design department in Malaysia was a partial exception. Forward integration by mining companies into smelting was less rare, but by no means the rule. Billiton's mining interests in Thailand were confined to one major offshore project and were not central to its major interest in smelting in Thailand and marketing. Its tin interests in Indonesia too were limited to a single mining project, and it had no part in Indonesian smelting. The development of local smelting in Bolivia and Indonesia was part of the increasing involvement of the state in those countries and is dealt with later in the chapter. In Malaysia one smelter, the Straits Trading Company, was owned by mainly Singaporean interests unconnected with mining (principally the Overseas Chinese Banking Corporation). The most interesting feature of smelting in the early 1970s was Patino's continuing interest after the nationalisation of its Bolivian mines in 1952. Patino's Consolidated Tin

Smelters merged in 1975 with Amalgamated Metal Corporation, a trading firm which the Patino group had controlled since 1968. As a result of the merger, AMC held controlling interests in the other Malaysian smelter, Datuk Keramat (formerly Eastern Smelting), the Nigerian smelter and Associated Tin Smelters in Australia. It also continued to control Williams, Harvey in England. AMC was taken over by the German mining multinational Preussag AG of Hanover in 1978. Preussag, though active in many non-ferrous metals, had no previous interests in tin, and this acquisition represented an essentially horizontal diversification. Apart from the former Patino connection with the London Tin Corporation's tin companies, AMC's mining interests were minimal. Its main associated companies were in trading, although it had a 20 per cent interest in a major German copper and lead refiner. In Malaysia, the other smelter, Straits Trading, had a number of investments in Malaysian tin mines, but those mines were not of great quantitative importance.

Over the 1970s then, the degree of vertical integration among the MNCs mining tin generally diminished. Patino's Amalgamated Metal Corporation which controlled half of Malaysia's smelting capacity was sold to a company with virtually no tin mining interests. The other Malaysian smelter remained outside the control of the mining industry altogether, and the Indonesian state mining corporation smelted almost the whole of that country's output. Bolivia too seemed at the time well on the way to state control over the smelting of domestic ore production. Billiton in Thailand was the main exception to this pattern, having a near monopoly of smelting and marketing, though on only one mining project. Rio Tinto Zinc wholly owned the Capper Pass smelter in England which smelted much of the Bolivian output that its domestic smelter could not handle, but RTZ's interests in tin mining were small (CIS, 1972). AMC's sale to Preussag also, of course, reduced vertical integration up to the marketing stage. Consolidated Goldfields controlled one of the largest metal traders, Tennant (active in Bolivia, for instance: see Widyono, 1977, p. 36), and from 1979 CGF owned smelting capacity via their Rooiberg company's new smelter in South Africa (*TI*, October 1979). Again, however, CGF were not one of the really major players in the world tin mining industry.

With regard to integration forward into the finished product there was little change over the 1970s since Fox (1974, p. 14) claimed there were virtually no links between tinplate producers and either smelters or miners. One exception was Billiton, which was a producer of solder. Another was Broken Hill Proprietory which operated large tinplating works in Australia and owned a small underground tin project in Indonesia.

The Development of the Brazilian Tin Industry

Throughout the 1960s and 1970s Brazil remained only a minor producer of tin. The focus of the industry during the 1960s and '70s shifted to the Amazon region, the area from which the massive expansions in output would come in the 1980s. Tin was developed in Rondonia in the Amazon in the 1960s by small-scale miners (the garimpeiros) who sold their output to companies who would smelt it. By 1962 Rondonia was producing more tin than any state in Brazil, though the country's total mine production was still under 1000 tonnes. In 1971 the Brazilian government banned garimpeiros from mining tin in Rondonia, and the mining was taken up by companies, some of whom had also been active in the region in the 1960s. In 1971 the first tin dredge was introduced into the Amazon.

More rapid development of the industry in the Amazon occurred in the latter half of the 1970s. Following the 1973 oil shock and the resulting balance of payments difficulties and accumulation of external debt, the Brazilian government took measures to give greater priority to mineral development generally, including various fiscal incentives. Foreign investment was encouraged, though companies with majority Brazilian ownership were given greater incentives (Lloyd and Wheeler, 1977). In the Amazon, companies were able also to benefit from special tax holidays designed to encourage the opening up the region. By the late 1970s the output of Brazil was greater than that of Nigeria or Zaire. Nevertheless, the development of the industry was inhibited by high infrastructural costs (e.g. for power supply), high labour costs (since labour had to be brought in), and high transport costs. Operating costs were cited as in the range $3 to $5 per cubic metre mined for most Brazilian mines (Engel, 1980, p. 13), whereas almost all South-East Asian tin mines had costs per cubic metre of well under $1. (Thoburn, 1981a, p. 124).

By the late 1970s the present industrial structure of Brazil's tin production was to a large extent in place. Four companies produced most of Rondonia's output, and that region accounted for three-quarters of Brazilian tin production. The two largest companies, Paranapanema and Brumadinho, were wholly Brazilian. The third largest, the Brascan group, was two-thirds owned by a Canadian investment company and one-third by Patino NV. Patino at that time also controlled the smelting company Cesbra, which in turn owned a mining venture, Mimbrasa. Mimbrasa prior to 1979 had been 50 per cent owned by the traders Phillip Brothers. A fifth mining group, owned by the Best smelter interests was planning a tin venture (Engel, 1980). Tin mining was closely integrated into smelting. The three largest smelters each having a link to one of the large mining companies.

Patino and Phillips had not been the only multinationals to take an interest in Brazilian tin. Billiton was said to have been active in tin exploration in Rondonia up to the beginning of the 1960s, after which it sold its interest to Brazilians, having failed to find adequate deposits. Also involved were the American firm W. R. Grace (with interests in medium-scale mining in Bolivia) and NL Industries (formerly the National Lead Company of the USA, which was a partner in Patino's early ventures).

Even by the late 1970s, however, the potential of Brazil was still unknown. Lloyd and Wheeler (1977, p. 57), while conceding that there were some claims that Brazil's tin reserves might amount to millions of tonnes, were of the view that the country's tin supply to maintain its domestic smelting industry was unlikely to come exclusively from local sources of cassiterite 'in the foreseeable future'. B. C. Engel, following his visit to Brazil for the International Tin Council in December 1979, reported (1980, p. 24) that:

> ... the general consensus of government and industry at the present time is that Brazil is unlikely to have an exportable surplus of much more than 2000 tonnes annually, and is not therefore expected to have a major impact on the world supply/demand equation.

PRODUCER POLICY

In Malaysia and Thailand taxation was used to secure for the host government a substantial share in the resource rents generated by tin, and there was little difference between the tax regimes facing local companies and those facing foreign firms. The use of taxation to change the retained value from current production is discussed in the next section, which also assesses other local gains from tin production in the main exporting countries. In addition to changing taxes, both Thailand and Malaysia in the 1970s sought directly to increase government ownership and control. Malaysia acquired control of the largest foreign-owned tin group, the London Tin Corporation, in a stock exchange takeover, and made foreign companies restructure their equity ownership. Thailand refused to renew several foreign mining leases and forced the restructuring of equity, as did the minor producers Zaire and Nigeria. Bolivia and Indonesia continued with the experiment of running a largely state-owned tin industry.

Policy Changes in Malaysia and Thailand

The Malaysian government's takeover of the London Tin Corporation was in line with the country's 'New Economic Policy' initiated in the *Second Malaysia Plan* (1971–5) after serious racial communal riots in 1969 had demonstrated a deep dissatisfaction among Malays with their economic position. Under the *Outline Perspective Plan*, which ran to 1990, it was intended that the economy would be restructured so that asset ownership would reflect more accurately the shares in population of the various ethnic groups. It was decided that foreigners should own only 30 per cent of the country's assets, Malays 30 per cent, and other Malaysians 40 per cent. The Malaysian portions would then represent more closely the shares of Malays and non-Malays in the total population.

The LTC takeover was achieved in a complicated series of manoeuvres[23] involving the Malaysian state investment corporation Pernas, MMC's future partner Charter Consolidated, Haw Par Brothers International in Singapore (the makers of Tiger Balm, a medicine which is a household name in Asia) and, for a time, the famous British property company Slater Walker.[24, 25] As a result, the newly-formed Malaysia Mining Corporation was able to control over 25 per cent of Malaysian output, which in turn was about two-thirds of the output of companies of foreign origin. MMC was managed by Pernas Charter Management, in which Charter and Pernas hold 50 per cent each. The formation of MMC left only the Osborne and Chappel group (in whose companies the smelter, the Straits Trading Company, had interests) and individual companies such as Conzinc Rio Tinto outside the government's control.

The formation of the Malaysia Mining Corporation was not a means of achieving total ownership, since LTC control (as mentioned earlier in the chapter) was secured through minority holdings, ownership of the other shares being widely dispersed. In 1977 the *Far Eastern Economic Review* (1 April 1977) estimated the overall Malaysian ownership of MMC companies at 45 per cent.

For Pernas, established under the New Economic Policy to hold assets in trust for indigenous Malays, the acquisition of the London Tin interests was an important means of moving its previously loss-making group into profitability. In the 1977–8 financial year Pernas recorded profits for the first time. In 1978 Pernas was absorbed into the PNB (Permodalan Nasional Bhd) investment trust with the continuing aim of acquiring assets for indigenous Malays (Elliott, 1989, pp. 37–8).

Malaysian ownership restructuring in tin was part of a well-articulated general policy towards foreign investment, the terms of which were made known well in advance. The initial moves towards restructuring in Thailand were a haphazard accompaniment to a change of government. In 1973 a military regime was overthrown in the context of student uprisings, and in the south of the country there was agitation against foreign ownership of natural resources. The most striking outcome was the revocation of the TEMCO lease in 1975, which had been owned by Union Carbide and Billiton, on the grounds that it had been obtained improperly by UC from the earlier government. The revocation of the lease was preceded by an almost explosive growth of illegal mining by small suction boats. According to miners in Phuket (interviewed in November 1978) both TEMCO and other foreign dredging companies had experienced difficulties with illegal miners before, but up to 1973 the companies had been able to enlist the support of the local police to drive them off. After the change of government, the local administration, particularly in Phanggna province, was sympathetic to the small boat owners, whose operations on foreign-owned leases were no longer checked.

Following the revocation of the TEMCO lease, and a change back to military government in 1976, other measures were taken both to decrease foreign ownership and to place it on a more certain basis. Existing mining companies were required to allow a majority Thai ownership 'at an appropriate time'. New offshore dredging ventures were still allowed, but had to have majority local ownership from the beginning, rising to 60 per cent within five years. Existing offshore dredging companies had to immediately conform to the 60 per cent share.[26] Aokam, which had been under the control of the Malaysia Mining Corporation, reconstituted itself as a 60 per cent Thai-owned Thai company – the Aow Kham Thai Company (*Bangkok Post*, 31 October 1978). Of the other members of the Malaysia Mining Corporation's TASK group in Thailand, Kamunting had its application for an inland lease refused and it left Thailand in 1975. Southern Kinta, particularly hard hit by illegal miners, shelved plans to construct a large-capacity shallow digging dredge for offshore work. Tongkah Harbour, the fourth TASK company (and the pioneer of dredging in South-East Asia) agreed at the end of 1979 to restructure to 60 per cent Thai ownership. The other group of foreign origin, Fairmont State, which worked inland, apparently had majority Thai ownership already (interview with Fairmont State, Bangkok, October 1979).

Billiton Thailand Ltd was formed after the end of TEMCO and the withdrawal of Union Carbide. In the late 1970s Billiton worked the old TEMCO lease as a contractor to the new leaseholder, the Offshore Mining Organisation (OMO), set up by the Thai government in 1975.

Although formed to control the former TEMCO lease, OMO was beginning in the late 1970s to branch into wider mining activity, and ordered a dredge of its own. A local Thai-Chinese dredging company with various links with the Malaysian Chinese mining community acquired secondhand dredges also to work as a contractor to OMO, and in 1978 another large local mining company (also with Malaysian links) started building a suction dredge also to work as an OMO contractor. The Thai restructuring seemed to have a more adverse effect on foreign investment than that in Malaysia, in spite of the fact that the maximum permissible foreign equity holding was smaller in Malaysia, while the taxation levels were similar. The TEMCO affair damaged confidence in the legal position of foreign investors, and in spite of the change of government the small boats were still poaching on foreign companies' legally-obtained leases. The boats apparently were backed by powerful local interests in southern Thailand. Some inland mines in the south had problems with communist insurgents and paid taxes to them. The fact that Thai output did not fall during the restructuring, except in 1975, is a reflection of the emergence of suction boat mining as the largest and fastest growing sector. The rapid rise in Thai output from 20,000 tonnes in 1974 to over 30,000 in 1978 also suggests that the boats were persuaded to sell their ore legally instead of smuggling it to Singapore. Phanggna province set up a buying agency, subcontracted to various individuals (Chulalongkorn University, 1976, pp. 29–30). Local miners, other than suction boat operators, seemed pessimistic about prospects when interviewed in 1978. Most reported they had diversified substantially in the last ten years into property, plantations and finance. Common complaints were the level of taxation, a level of official concern with environmental protection which the miners considered excessive, and a difficulty in obtaining bank finance.

In Malaysia the MMC takeover was accompanied by a large fall in output (from around 75,000 tonnes at the beginning of the 1970s to a low of just under 59,000 tonnes in 1977), though the declining trend was halted in 1978. Undoubtedly the restructuring of ownership, under which foreign companies could hold normally only 30 per cent of the equity, acted somewhat to discourage new projects. Complaints also were made by all sectors of the industry in Malaysia about 'penal' taxation. Although dredging output failed to expand at a time of extremely high prices in the 1970s, the contraction of output was due mainly to a decline in the gravel pumping sector. This was partly due to a taxation structure which affected this sector particularly adversely, but more due to difficulty in securing leases and finding new mining land (Thoburn, 1977b).

The lack of prospecting in the 1970s in Malaysia and the consequent shortage of new land can be blamed partly on the New Economic Policy (NEP) and on taxation. However, leases were difficult to secure no matter how strong a company's willingness to invest. This problem was overwhelmingly the result of a conflict of interest between the federal government and the governments of the individual states. In the Malaysian federation the latter had the virtually exclusive right to grant new leases, yet they received only 10 per cent of the tin export duty, none of the other taxes on mining, and only a small land premium. Thus their incentive to alienate mining land was slight. The two west coast states of Perak and Selangor produced about 90 per cent of the country's tin output, and Selangor took the lead in a tough policy to secure additional gains for its state government. Initially, even the Malaysia Mining Corporation had lease applications refused in Selangor, apparently on the grounds that its operating companies did not conform to the NEP ownership guidelines.

The vehicle for greater Selangor state government participation in mining, since it did not have the power to tax, became the Selangor State corporation, Kumpulan Perangsang Selangor (KPS). After 1978, the declared state government policy became that no new leases over 500 acres (effectively any dredging lease) would be granted solely to the private sector. Instead, KPS would hold a 70 per cent stake in foreign-owned ventures and 51 per cent with local companies,[27] and these conditions would apply to the renewal of existing leases too. In practice the guidelines seem to have been applied with some flexibility, although the distinction between 'local' and 'foreign' was sometimes unclear. State Development Corporations (SDCs) in other states also entered into joint ventures, though generally on less stringent terms than in Selangor. Joint ventures with obviously foreign companies had to conform to the '70/30' Foreign Investment Committee Guidelines.

Malaysia Mining Corporation was also a prospective joint venture partner, and Selangor state insisted on treating the MMC's operating companies as foreign, stipulating a 70 per cent KPS stake even before leases were renewed. In the huge new Kuala Langat deposit in Selangor with reserves estimated as equivalent to about five years' total Malaysian output, MMC was able to secure only a 30 per cent share; and this after protracted negotiations and the cancellation of a previous agreement (TI, December 1979).

The joint venture system involving state governments appeared to resolve the problem of land leases for dredging, both for foreign companies and the MMC. Since there was significant foreign ownership in many of MMC's operating companies, the increased participation by state governments and Malay companies resulted in greater local ownership, though the shares in Malay companies were not widely dispersed among the Malay population. For local Chinese miners the pressure to restructure in favour of Malays was more subtle. Officially, even in Selangor there was no change of policy towards local companies requiring lease renewal (or new leases) of under 500 acres. In practice there may have been some preference given to bumiputra companies for new leases. When interviewed in 1978 the Bumiputra Chamber of Mines had about 120 members (there were then about 800 gravel pump mines in Malaysia), but roughly half had Chinese partners.

By the late 1970s, Malay participation in the mining labour force had increased to over a quarter, compared with about 5 per cent in the colonial past (Thoburn, 1977b, p. 92), though some state government enterprises such as Timah Langat employed a wholly Malay workforce. Many mines preferred an ethnic mixture, however, since the mine could then be kept open through the country's many festivals.

In Thailand no moves to increase government ownership and control in smelting and marketing occurred. The Thaisarco smelter remained 100 per cent owned by Billiton, which remained 100 per cent owned by Royal Dutch Shell, and Billiton marketed Thaisarco's entire output. Unlike Malaysia, the smelter did not have its own network of dealers to buy from small miners, and ore dealing was predominantly in the hands of the larger gravel pump miners together with the representatives of the provincial organisation which bought from the small boat operators. Some smuggling continued and the smelters in Singapore (a country which itself produces virtually no tin) were producing annually about 4000 tonnes of tin (*TI*, May 1980), mostly, it is thought, from Thailand and Indonesia. The establishment of Thaisarco in 1965 was supported by Thailand's Board of Investment with a prohibition on exports of unsmelted ore (which mostly had previously been sent to Malaysia) and a five-year ban on other smelters being established. In 1976, permission was given by the Board for two new smelters to be established in the region near Bangkok to deal with ore from the central and northern regions of the country (Sriwichchar and Pithaya, 1977). Of these two, the Thai Present Company was 90 per cent foreign-owned. The Thai Pioneer Company, which was Thai-owned, had links with the German metal dealing firm Metallgesellschaft which would have supplied smelting expertise and dealt with marketing. In 1980 the planned location of the Thai Present smelter was switched to Phuket, where it was to be in competition with Thaisacro (*TI*, May 1980).

The Thaisarco smelter charged according to the same scale as the smelters in Penang. Billiton (interview, June 1978) claimed that it was smelting rather than marketing, which was the profitable part of the operation. In Malaysian smelting there is a long history of high profitability (Thoburn, 1977b, p. 127).

By the end of the 1970s, negotiations were starting for the Malaysia Mining Corporation to acquire an equity share in each of the two smelters. As a bargaining move, MMC opened bids for the construction of a third smelter to handle its entire output. In 1981 MMC acquired a substantial stake in the Straits Trading Company's smelter. A new firm, the Malaysia Smelting Corporation was set up to run the smelter, with MMC owning 42 per cent and the Straits Trading Company 58 per cent. In 1981, too, some links were formed with the Datuk Keramat smelter.[28]

In any case, MMC dealt with its own marketing after 1978, when it started to toll-smelt its output with the local smelters, and cooperate in marketing with Anglo–Chemical & Ore in London (*TI*, January and May 1980), a subsidiary of the large metal trading firm Phillips (also active in Bolivia). Previously all ore had been sold on the Penang market on an arms-length basis, with the result that miners could not keep control of their ore through the smelter. This may well have been a factor too in the lack of integration backwards into tin mining by steel companies owning tinplate capacity. Bethlehem Steel for example, when interviewed in Bangkok in autumn 1978, mentioned this as a difficulty in securing a captive source of tin.

The bypassing of the Penang market by MMC (and later also by the Selangor state mining organisation KPS) was described by the Malaysian Minister of Primary Industries at the time as an anticipation of the establishment of a Kuala Lumpur metal exchange (*TI*, January 1979). The Kuala Lumpur Commodities Exchange (KLCE) opened in 1980. The KLCE's physical market for tin, the Kuala Lumpur Tin Market, opened in October 1984, replacing Penang. The aim was not only to replace the physical sales previously centred on the Penang market, but also to move a proportion of 'paper' trade (in tin futures) away from the New York market and the London Metal Exchange.

The Operation of State-owned Enterprises in Bolivia and Indonesia
Bolivia

After nearly a decade of falling production, in 1961 the Bolivian government introduced the so-called Triangular Operation to rehabilitate Comibol. Finance was provided by three donors (hence 'triangular'): the US government, the German government and the Inter-American Development Bank. The plan provided loans for new exploration, re-equipment and to aid the reduction of surplus labour. Credits were also secured from Argentina and from large ore buyers (Zondag, 1966, ch. 19). In addition, the MNR government having strengthened its power base among the peasants (agriculture employed the bulk of the Bolivian workforce) felt able to take a harder line about the mineworkers, a line harshly strengthened after the takeover of power by a military government in 1964. Little, it seems, was achieved by the plan. Most of the loan was spent on immediately-needed supplies rather than on long-term investment. There was virtually no provision for strengthening Comibol's internal organisation once the plan's foreign experts left, and the measures were sharply opposed by the mineworkers (Ayub and Hashimoto, 1985, pp. 15–18)

In the 1970s, with a rising trend of world tin prices particularly after 1973, there was a recovery in Comibol's production, aided by devaluation in 1972 and by a period of labour peace in the mid-1970s. There was also an improved supply of external credit reflecting both the excess supply of loanable funds in the world economy (eventually to lead to the world debt crisis in the early 1980s), and Bolivia's potential as an oil exporter (Ayub and Hashimoto, 1985, pp. 18–19). From 1971 to 1977 annual average price always yielded a before-tax profit for Comibol,[29] but royalty payments (a tax on output imposed largely regardless of profitability) creamed off the bulk of the surplus and led to financial losses in 1976 and 1978. In 1978, royalty payments turned a 22 per cent profit on turnover into a 2 per cent loss, and this was a figure which included the underground operations of the more profitable medium mining sector. In 1976 royalty payments were equivalent to 21 per cent of the average LME price (and taxes to a further 8 per cent), and in their absence Comibol would have made an average 18 per cent profit on turnover. In 1979 when the real tin price was exceptionally high (see Figure 1.1), Comibol earned a financial profit, though not in 1980, and by 1981 the operating surplus itself had ceased to be positive (Ayub and Hashimoto, 1985, p. 45).

Heavy taxation of a state enterprise is not in itself necessarily undesirable. The state may well put the profits to better use than the enterprise itself, but in Comibol's case it seems that taxation was so heavy that opportunities for essential re-equipment and exploration were lost. The Harvard team which advised the Bolivian government on its mining taxation policy in the mid-1970s (see Gillis *et al.*, 1978) concluded that over the ten years to 1975 only about $10 million had been spent on exploration[30] and only one significant new deposit had been found. Gillis *et al.* also stressed a lack of fully-effective management control, and noted the trade off between higher accounting profitability and the social responsibility undertaken in employing a larger labour force than technically necessary. In the late 1970s various new projects were announced, including the National Mineral Exploration Fund set up in 1979 with International Development Agency finance (*TI*, August 1979) and a dozen or so new projects (mostly Comibol's) were said to be under consideration, with outside finance being sought from the World Bank and other agencies. Yet in 1979 the resigning general manager of Comibol complained about the outdated equipment from which the corporation suffered and described Comibol as being on the 'verge of bankruptcy' (*TI*, April and October 1979).

Jordan and Warhurst (1992) argue that the 1970s period of high tin prices (and more readily available international loan capital) simply masked Comibol's longer term difficulties. They see these as springing from a mode of corporate behaviour in which the maintenance of output took precedence over the urgent need to innovate (in contrast to many of Comibol's competitors). There was a strong social dynamic protecting excessive employment, but not providing workers with incentives or good facilities, and an organisational structure under which a 'puppet management' installed by the state was unable to exercise effective control.

In addition to Comibol's activities, about a third of the country's tin output in the 1970s was being produced by privately-owned mines. These were officially classified into a 'medium' and a 'small' scale sector. The medium sector, consisting of nearly thirty firms, employed about 8000 workers and all of its activities were underground except for the operation of two dredges. According to Gillis *et al.* (1978, p. 32) only two firms were wholly foreign-owned, and four or five had minority foreign participation, though these figures are not broken down to allow the importance of foreign ownership in tin production to be assessed. Some medium mines were important producers of tungsten and antimony too, and six mines in fact produced 70 per cent of the sector's tin output. Fox (1974, p. 66) estimated that in 1966 40 per cent the sector's output came from wholly Bolivian-owned and under 10 per cent from foreign owned mines,[31] Chilean and American capital being of some importance. Small mines employed a further 3000–4000, and worker cooperatives about another 20,000, though the latter's output was small, much of the small-scale mining being on a seasonal basis.

In 1966–9 some reinvestment occurred in medium mines resulting in updating of their equipment, but little further investment had taken place when the Harvard team visited Bolivia. *Tin International* (October 1979) reported investment programmes of $17 million and $22 million in 1977–8 but the medium mines were complaining strongly about the tax system based on royalties (which made no allowance for production cost differences) and they had tended to reinvest in agricultural ventures (where tax provisions were more favourable) rather than in mining. Medium mines, however, in many cases had lower costs than Comibol, basically because of higher labour productivity.

Indonesia

The early 1960s were a time of general economic dislocation under Sukarno's 'guided economy'. In 1965, just before a military coup established General Suharto as the new president, inflation was running at 650 per cent per annum. An inappropriately high currency exchange rate was maintained, reducing the industry's profits in a situation where it had little influence on price, the divergence between the official and the black market exchange rate being about 500 per cent (Mackie, 1971, p. 64). A serious problem too was the previous lack of investment. Although six new dredges had been commissioned between 1947 and 1948 the bulk of the dredge fleet was of an advanced age (P. N. Timah, 1972). Following the military coup in 1966, a rehabilitation programme was begun with Dutch aid. In 1966 a new dredge was commissioned to work offshore at Bangka (Fox, 1974, p. 36). After a trough in output of 12–13,000 tonnes in 1966–7, during which time many Chinese workers were expelled from the tin islands, production rose steadily, reaching over 25,000 tonnes by the mid-1970s.

In the 1970s, P. T. Timah embarked on a programme of expansion based on seagoing dredges. Three new large capacity dredges, designed to produce 1000 to 2000 tonnes per year, formed an important part of the planned increase to 35,000 tonnes by 1983 under the third five-year plan, Repelita III (1979–83) (*TI*, August 1979). Equally important was a gradual replacement of the existing dredge fleet, two-thirds of the capacity of which was of pre-war construction (P. N. Timah, 1972). A great deal of rehabilitation of the older dredges was also undertaken, including the conversion of all the steam-driven dredges to electric or diesel-electric power.

Rehabilitation was internally financed by Timah, and taxation at the end of the 1960s was reduced allow this. In 1968 and 1969 Timah was allowed to use part of profit before tax to finance rehabilitation expenditure.[32] Unlike Comibol, Timah's basically more profitable operation and lower tax levels allowed it to continue to finance rehabilitation and new investment internally. The change in corporate status to P. T. Tambang Timah Persero in 1976 gave Timah the power to raise money on the open market. The change from 'P. N.' to 'P. T. Persero' indicated that it was no longer a State enterprise, but a private limited company (albeit with the government as sole share-owner). After-tax profit in 1977 was equivalent to US $39 million (on a turnover of $250 million), which was enough by itself to have financed an entire new sea dredge project. In practice the shareholders (the government) decided on the distribution of profits,[33] but investment plans were submitted to the investment coordinating body, BKPM (Badan Koordinasi Penanaman Modal), which allowed offsets against tax. Timah's investment expenditure was substantial in the 1970s, peaking at 28 per cent of the value of output in 1973 (Thoburn, 1981a, p. 78).

The use of a light export tax (10 per cent of the value of tin metal sales) and a substantial tax on Timah's profit (at 45 per cent) was in accordance with the generally-accepted principles of mining taxation (e.g. Gillis *et al.*, 1978). In economic theory, tax could be levied entirely on profits. In practice this gives greater scope for evasion or, in a state company, for the inflation of costs by, for example, excessive staff benefits. There was no provision for taxes to rise if the tin price did so (in contrast to the third generation contract applied to new foreign mining investment in Indonesia). The system was considerably more efficient than the Bolivian export royalty system in the sense of allowing adequate resources for new and replacement investment and exploration while also generating substantial tax revenue. One effect of investment and rehabilitation was a considerable rise in labour productivity. Output per worker per year rose from 0.5 tonnes in 1966 to roughly 0.9 in 1975.[34] This compares with 1.6 tonnes in Malaysia in 1975 where grades were *lower* (Malaysia, *BSMI*, 1975). However, the figures are not strictly comparable since Timah was far more vertically-integrated than the Malaysian tin industry, supplying its own power and repair facilities whereas the Malaysian mines bought such services from outside the industry, and workers in the supplying industries are not included in the

Malaysian statistics. Timah also regarded itself as having social obligations to employ a workforce larger than necessary on strictly commercial grounds. It is also worth noting that mining costs per tonne of concentrates differed substantially between the tin islands. Bangka produced 69 per cent of 1977 output, but accounted for only 53 per cent of Timah's mining costs,[35] the difference being primarily a reflection of Bangka's higher ore grades.

Although tin was Indonesia's largest mineral export after petroleum, Timah occupied a far less important position in the mining economy than did Comibol in Bolivia. Whereas Comibol also accounted for large shares in Bolivian production of other minerals (excluding oil),[36] Timah had the power to diversify only after its change of status in 1976, and moved only to a small extent into kaolin, bricks and some minor items. Other state enterprises in Indonesia were responsible for other mineral production: Pertamina for oil, P. N. Batubara for coal and P. T. Aneka Tambang for other minerals.

Though Timah produced about 90 per cent of Indonesia's tin output, not all of its production was from its own facilities. Gravel pumps produced nearly half of total output and slightly more than half of these mines were operated by contractors. Although in the small project of Bankinang in Sumatra the contractor supplied his own equipment and fuel, and sold the ore to Timah at a fixed price, in Bangka and Belitung the contractors normally supplied only labour. Presumably the system gave the corporation some flexibility in altering the effective size of its labour force. It may also have been a means of using the mining expertise of the Chinese community who were excluded from direct employment.[37] Since the payments to contractors appeared not to vary, at least in the short term, with the tin price, there may have been some incentive to smuggling at times of high prices. Early in 1979 there was a major drive against smuggling from Bangka to Singapore.

The Development of Smelter Capacity in Bolivia and Indonesia

Before the Indonesian and Bolivian nationalisations took place, the bulk of each country's tin concentrates was shipped overseas for smelting. In the Indonesian case they were sent to the Netherlands until 1959 and then to Malaysia until a confrontation in 1963 (Fryer and Jackson, 1977, p. 195), though before the second world war there had been some local smelting. From Bolivia concentrates were mainly sent to Europe, where the Patino organisation owned the largest smelter, Williams, Harvey. After nationalisation both countries integrated forward into smelting and the Indonesians established a fully-integrated organisation through to marketing of the metal.

The smelting of alluvial tin concentrates, in Malaysia at least, has been an extremely profitable occupation over the years[38] and this seems to be true of Thailand too.[39] In Indonesia, producing basically similar concentrates, considerable difficulties were experienced in the early stages of establishing a smelter, which was designed and built over the period 1961 to 1967 by a German company. The major problem, apparently, was that the smelter was of an essentially experimental design (using rotary furnaces instead of the more usual stationary ones) and by 1970 had achieved less than two-thirds of rated capacity. Additional, more conventional, capacity designed in Denmark and built by a Singapore firm, was installed during 1973–5 and gave Indonesia self sufficiency. All concentrates by the late 1970s were being locally-smelted except for those from the single (foreign-owned) underground mine, which were low grade and were sent to Malaysia. In 1979 the smelting unit started building a further furnace to raise total capacity to 39,000 tonnes per year[40] to deal with increases in future output.

According to Timah's director of marketing in 1973 (ITC, 1974, p. 15) the smelting operations already 'made money', though in 1978 costs appeared to be higher than the charges levied by Malaysian smelters. Whether or not some cost saving could have been achieved by continuing to ship concentrates to Malaysia, Indonesia's doing its own smelting permitted Timah to market virtually the whole of Indonesia's output, much of it on long-term contract.

The establishment of local smelting capacity took considerably longer in Bolivia, and in 1979 40 per cent was still being sent abroad. Smelting the complex hard-rock ores of Bolivia poses more problems than are found with the purer alluvial concentrates of Asia.[41] Bolivia has also been hampered by the lack of indigenous coal or hydro-electric power and by the power loss which occurs in smelting at Bolivia's high altitudes (Ayub and Hashimoto, 1985, p. 55). In Asia most mines concentrate their tin-bearing ore to about 70–5 per cent purity. In Bolivia the purity fell from about 60 per cent in the 1940s to 35–50 per cent by the late 1970s, and many small mines delivered concentrate of below 20 per cent (Gillis *et al.*, 1978, pp. 34–6). The Bolivian smelting organisation ENAFBOL (*Empresa Nacional de Fundiciones de Bolivia*) was organisationally separate from Comibol and had been operating a smelter since 1971, built by the same German company which constructed the first Indonesian smelter. Costs were high in the beginning because infrastructure was built for a planned 20,000 tonnes per year capacity whereas only 6,000 tonnes was built initially, though it had risen to 11,000 by 1976 (Gillis *et al.*, 1978, pp. 34–5). The ENAFBOL smelter could only handle the higher grade concentrates, which were previously shipped for smelting to Williams, Harvey in the UK. In 1979 the building of a low-grade smelter was started together with the expansion of the main smelter (at Vinto) to 20,000 tonnes in an attempt to give Bolivia self-sufficiency (TI, April 1979). The low grade smelter was completed in 1980 with 10,000 tonnes capacity, handling the concentrates previously sent to Capper Pass (owned by Rio Tinto Zinc) in the UK and Metallgesellschaft (Ayub and Hashimoto, 1985, p. 56).

Whereas the Indonesians appear satisfied with the performance of their smelter, the outgoing manager of Comibol complained in 1979 that he could make another $20 million a year if freed from the obligation to sell his concentrates to ENAFBOL (*TI*, October 1979). This alleged loss represented 7 per cent of Comibol's turnover, while in Malaysia smelting normally cost less than 2 per cent of the value of output. The Bolivian operation suffered apparently from a lack of suitable fuel and lack of higher grade concentrates to achieve the suitable mix than overseas smelters obtained (Zondag, 1966, p. 97). Nor did the Bolivians successfully enter marketing in the Indonesian fashion. In 1973 some 70 per cent of medium mines sales were sold directly to foreign smelters, though some were to the mining bank BAMIN (Banco Minero) which also bought the output of the small mines. According to Widyono (1977, pp. 33–6), international metal trading companies were still very active in Bolivia, with commissions up to 20 per cent, the main firms being Phillip Brothers, Tennants, Metal Traders and a subsidiary of the Amalgamated Metal Corporation. When Ayub and Hashimoto (1985) reported on the Bolivian tin industry in the early 1980s for the World Bank, they commented that (p. 57):

> On all accounts, the performance of ENAF has been one of unmitigated failure and a good example of how the achievement of a set of reasonable and realistic objectives can be frustrated by mismanagement, poor planning, lack of coordination, and technical and industrial relations problems.

Policies in Minor Producers

Nigeria and Zaire in the 1970s took measures similar to those of Malaysia and Thailand to increase local ownership. Under the Nigerian Enterprises Promotion Decree first enacted in 1972, foreign companies by the end of 1978 were expected to have sold 60 per cent of their equity to Nigerians. The largest foreign company, Amalgamated Tin Mines of Nigeria (at the time a subsidiary of the Malaysia Mining Corporation, who had taken it over from the London Tin Corporation) noted in its 1978 annual report that its shares had been sold mainly to a state body, the Nigerian Mining Corporation. Shares were sold at a price decided by the Nigerian Capital Issues Committee. Similar sales have been made by other large companies, although there were complaints about the need to sell shares on a saturated market (*TI*, December 1978, February and December 1979). The Nigerian Mining Corporation also took a 21 per cent interest in the Makeri smelter (Engel and Allen, 1979, p. 24), set up in 1962 by Consolidated Tin

Smelters, which had quickly pushed out of business a smelter set up shortly before by Portuguese interests (Freund, 1981, pp. 215–17).

In Zaire, the Sominki company was formed in 1976 as an amalgamation of three smaller mining groups, with 28 per cent of the equity held by the government of Zaire (*TI*, December 1979). The only other foreign mining company in tin in Zaire, Geomines, also of Belgian origin, was reformed in 1966 into Zairetain, in which the Zaire government had 50 per cent. In Ruanda, the Somira company was established in 1973 to include a number of the major tin companies, with 49 per cent owned by the government (*TI*, November 1978). Australia required 50 per cent local ownership, but this was administered flexibly (Engel and Allen, 1979, p. 153).

Taxation levels varied considerably between the minor producers as far as export duty is concerned, the rates for Nigeria being 16 per cent regardless of price, 2–4 per cent for Australia (varying with state) and zero for Zaire. Corporation tax was levied in those three countries at 45 per cent, 46 per cent and 50 per cent, respectively. In the UK no export duty was payable and corporation tax varied from 42–52 per cent according to the size of profits (Engel and Allen, 1979, ch. 4).

In Zaire and Nigeria output declined markedly. In 1978 Zaire's output was little more than half that of the late 1960s, and that of Nigeria was little more than a third. Nigeria experienced a great outflow of foreign personnel after its takeover of tin mining. Engel and Allen (1979, p. 24) attribute the Nigerian decline in part to falling ore grades and a lack of past exploration, but the adverse effects of oil expansion on the rest of the economy also operated against tin (Freund, 1981, pp. 226–7).

THE DEVELOPMENT GAINS FROM TIN EXPORTING

In the literature on the benefits to the host-country from mineral development, introduced in Chapter 1, analysis centred on the measuring of retained value (RV), on measuring the country's share of final consumer value (FCV), and on wider development effects such as local market creation, income distribution changes and employment creation. As noted there, such measures can only be used by themselves to summarise mineral export development effects if tin export activity is 'marginal' to the economy.[42] If it is not marginal one would expect the equilibrium real exchange rate in particular to differ from what it would be in the absence of the mineral export activity. In such circumstances, there will be benefits and costs in other sectors beyond those identified through the analysis of linkages which is conducted as part of the measurement of retained value and the distribution of final consumer value.

Of the major tin producers, only in Bolivia is tin mining clearly nonmarginal. In the 1970s tin in Bolivia generated 57 per cent of export income compared with 11 per cent in Malaysia, 7 per cent in Thailand and 2 per cent in Indonesia (1977 figures). Thus only Bolivia would be classified as an LDC 'mineral export economy' on Nankani's (1979, 1980) definition (see Chapter 1), having distinct characteristics, problems and prospects compared with other primary exporters.[43] Indonesia and Nigeria could have been considered on account of their petroleum, and Zaire for its copper, but not for tin. The first subsection here, which discusses non-marginal development effects, concentrates on Bolivia, and also gives a brief account of tin-mining and its effects on income distribution in the other main producers.

Exchange Rate Changes and Other Non-marginal Effects

The Harvard team in Bolivia in 1973–4 measured the mining sector's non-marginal effects (Gillis *et al.*, 1978, pp. 82–9), and their empirical results are summarised here. In the absence of the large non-fuel mining sector (predominantly tin-mining) the exchange rate would need to depreciate. As a result of the depreciation the local currency price of non-mining exports would rise,[44] and their supply would increase, while the local currency rise in import prices would reduce the quantity of imports purchased. The resulting change in welfare is the saving in resource costs resulting from the absence of mining activity less the additional costs of increasing the supply of other exports, and less the welfare costs of the change in expenditure and fall in consumer surplus of the reduced import volume. Using linear demand and supply curves and making simple assumptions about the price elasticities of demand for imports and supply of non-mining exports, a new and lower equilibrium exchange is calculated, and the consequent resource cost and welfare changes as the economy moves to this rate are estimated. The loss to the economy which would result if the non-fuel mining sector were absent is estimated by the Gillis team to be equal to approximately to a third of the net foreign exchange (export revenue less directly imported inputs) generated by the mineral exports. This gain is additional to the RV figure calculated by the team on the basis of the share of export revenue spent on locally-produced products out of *increases* in mineral output or price.[45]

The boost to the exchange rate from the growth of a large mineral exporting sector is the way in which additional imports are made available to the economy at lower real cost in terms of domestic resources. The higher exchange rate, however, damages the prospects of non-mineral exports and of import competing industries (though in the latter's case there is some offsetting benefit as demand for most commodities rises with the economy's increase in real income). If the exchange rate appreciation were the result of an expansion of agricultural or manufactured exports, whose existence might be expected to continue indefinitely, the damage to other exports and to other forms of domestic economic activity might not be a matter of concern. Even in this case, though, a high degree of specialisation would leave the economy vulnerable to shifts in export demand, and there would be problems if the export industries did create sufficient employment. Because mineral production is subject to exhaustion, there is likely to come a time when exports will be needed to provide foreign exchange and when import competing industries also will need to expand; both requiring a depreciation of the exchange rate. Of course, in static reallocative terms this should not be a problem. In the context of long-term development, however, there is a danger that domestic industries may be forced out of existence by imports while agricultural and other export activities may never develop to a stage where they can compete on world markets. Thus, a free foreign exchange market may fail to provide efficient intertemporal price signals, and the exchange rate may settle at a level which would be approximate only if the current sales of mineral exports were to continue forever. Also, a dominant minerals sector may not be able to provide sufficient employment, especially if a dualistic wage structure results from mining.

Bolivia certainly had not succeeded in becoming an exporter of manufactured products. In 1975, 96 per cent of Bolivia's exports were of primary commodities (World Bank, 1980b). In 1963 agricultural products were only 4.6 per cent of total exports, the rest being minerals (principally tin and petroleum) though even this represented agricultural export growth compared with the previous decade (Zondag, 1966, p. 177). Adverse effects on domestic agriculture via food imports have to some extent been counteracted by policies followed since the political upheaval of 1952, which led to a stress on agricultural development (especially by new settlement in the east and encouraged by transport improvements) and on petroleum exporting. Indeed, tin-mining was deprived of necessary reinvestment funds in order to finance government expenditure elsewhere. Nankani's (1979) survey of mineral exporters noted that (like many other mineral export economies) Bolivia suffered from a relatively low savings ratio, but he stressed its high agricultural potential. Inflation was high and was caused to large extent by government deficit financing during slumps in mineral prices. On the other hand Bolivia, unlike many other mineral export economies, did not have any significant differential of mining wages relative to those in manufacturing, nor was the manufacturing wage to GNP per head ratio high compared with most LDCs. Thus Bolivia did not have a dual wage structure in favour of mining. In fact, Bolivian mineworkers' living standards have been very low.

Since it is being argued that the tin industries in Malaysia, Thailand and Indonesia were 'marginal' to the economy in terms of their effect on the exchange rate, one would also not expect their effect on the general distribution of income in the economy to be significant. In any case in Malaysia (where tin had the largest share of exports among South-East Asian producers in the 1970s) mining wages differentials in the 1970s were not particularly high in relation to those in manufacturing (see later). Also, in Thailand and (especially) in Indonesia tin-mining is geographically distant from the main centres of economic activity and one would expect its influence to be diminished. Finally, it can be noted that in the more distant past in Malaysia tin was not a marginal activity, and one of its effects (and in Thailand and Indonesia too) was to induce a large immigration of Chinese workers. However, the subsequent role of the overseas Chinese in South-East Asian development is certainly beyond the scope of this study. In general, there is a problem in estimating the non-marginal effects of an export activity which changes the social structure of an economy considerably. For example it would seem unreal to describe the 'without-rubber' situation in Malaysia, where massive importation of Indian (and Chinese) workers occurred and where the economy of Malay village society was radically altered in much of the country.

Retained Value

Retained value is still widely used (with some reservations, to be discussed later) as an indication of the distribution of gains between a host-country and foreign mining companies. RV shows that the extent to which a foreign-owned activity has remained an 'enclave' – and it is a convenient method of assembling data on local payments made out of export income – whose subsequent development effects can be traced. An increase in RV which results from a policy of integration backwards into intermediate products for the industry is subject to the proviso that the production should not be 'inappropriate' for local conditions. That issue is discussed in the next subsection, on linkages, though in fact many tin mining linkages have not been a result of government policy. Increases in RV resulting from increased taxes or from equity restructuring sometimes provide only short-term gains, since they may damage foreigners' incentives to invest. However, taxation on tin-mining in the 1970s, at least in South-East Asia, did not appear to have been excessive, despite some mining industry claims to the contrary (Thoburn, 1978b; Thoburn, 1981a, ch. 6). RV also is likely to vary between different production techniques, often associated in tin with different ownership. Thus encouragement to a locally-owned sector such as gravel pumping or suction boat mining is sometimes an alternative or a supplement to restructuring the payments of a foreign-owned sector to raise RV.

Table 5.3 shows estimates of RV for Malaysia, Thailand and Indonesia.[46] In Malaysian dredging in 1978, RV was nearly 77 per cent (which was a rise

TABLE 5.3   The distribution of the gains from tin mining, 1978

	Gross profit	Retained value	Net economic gain	Net economic gain after profit outflow
Malaysia				
Dredging	65.9	76.8	60.7	52.6
Gravel pumping	43.3	86.0	61.0	62.0*
			41.7	41.077**
Thailand				
Offshore dredging	68.7	58.8	NA	NA
Onshore dredging	63.1	86.3	34.6	30.2
Gravel pumping	54.5	79.1	51.0	51.0
Suction boats	54.7	NA	NA	NA
Indonesia				
Offshore dredging	61.9	93.3	56.7	6.7
Onshore dredging	44.3	90.9	NA	NA
Gravel pumping	10.3	83.8	15.9	15.9

* (large mines)
** (small mines)
Source: Thoburn (1981a, p. 163)

from approximately 72 per cent in 1972). Wage and salary payments (about 8 per cent) were a lower proportion of output value than in the early 1970s and the 1960s, when they averaged 11–15 per cent.[47] They were about 90 per cent local, reflecting the fact that Malaysians held all but a very few of the most senior positions. The most striking feature is the size of gross profits (nearly 66 per cent), reflecting the highly intra-marginal nature of Malaysian tin production and the resource rents generated thereby with a high real tin price. Though the late 1970s was a time of higher real tin prices than 1972, in 1972 the figure was still over 50 per cent and gross profits of nearly 70 per cent were also recorded in the 1960s. The increased local ownership resulting from Malaysia's takeover of London Tin, discussed earlier in this chapter, resulted in a larger share of after-tax profits remaining in Malaysia compared with the early 1970s, but to a large extent the high overall RV figure still derived from tax. Export duty rose from about 16–17 per cent in the mid-1970s to nearly 29 per cent as tin prices rose, and the duty structure was made more progressive. Income tax was first imposed on companies in 1948 at a rate of 20 per cent and was increased to 30 per cent in 1951 and to 40 per cent in 1959 (Lim Chong Yah, 1967, p. 265), and remained at that level in the 1970s. In addition a variable tin profits tax was levied from 1965,[48] and a 5 per cent development tax from 1967. The export duty dates from the nineteenth century and in most post-war years it took about 15 per cent of output value.

Compared with the period before the second world war, government policies, together with increased local production of inputs, brought about a substantial change in the distribution of gains. In 1928, the first year for which sectoral output statistics are available for Malaya, wages were less than 10 per cent of gross output value, export duty 14 per cent and other operating costs probably about 10 per cent. Given the higher foreign content of wages and salaries, and the remission of profits overseas, RV could have been as little as 25–30 per cent.

The payments structure of Thai and Malaysian dredging operations was broadly similar, though direct tax in Thailand was lower and export royalty higher. In onshore dredging in Thailand profits were slightly lower reflecting lower grades and older equipment. RV in offshore dredging in Thailand was slightly lower than that in dredging in Malaysia. The difference is partly that the Thai operations used imported diesel as fuel instead of local electricity,[49] but more because the table assumes that in 1978 local ownership was nil, prior to the late 1978 restructuring. Restructuring the equity to give 60 per cent Thai ownership would raise the RV figure almost to the Malaysian level. RV in inland dredging was much higher in Thailand than for the offshore operations, partly because it was more labour-intensive and because it tended to consume local electricity, but also because it mainly was locally-owned.

Comparing dredging in the two countries to gravel pumping indicates that in Malaysia gravel pumping had the higher RV; and in Thailand RV was higher in onshore (i.e. inland) dredging than in gravel pumping, though both had higher RV than offshore dredging. The superiority of Thai onshore dredging to gravel pumping in terms of RV was largely because gravel pumping consumed imported diesel oil in large quantities while paying only a slightly higher share of its output value to local labour. RV in the suction boat sector, which was booming in the 1970s, is an estimate based on data from a small sample survey by a team from Chulalongkorn University (1976). Of the boats' cost items, only petroleum products, some of their repair components and a few minor items will have been imported, and it is difficult to see how RV could have been much less than 90 per cent of current revenue. Of course, RV would have been lower for boats which smuggled their ore out of Thailand!

In all sectors in Malaysia and Thailand the high profitability before tax in the 1970s is clear. In Thailand, more reliance was placed on export royalty than in Malaysia, and as in Malaysia this export taxation dated back to the last century. Taxes on company profits in Thailand dated from before the second world war, but there was a serious problem of non-compliance in the corporate sector generally (Ingram, 1971, pp. 184–7, 299–301) so the reliance on a more easily collected output tax is understandable. High before-tax profitability in dredging was also a feature of Indonesian tin-mining, especially offshore where grades were high. RV in Indonesian dredging of over 90 per cent reflects mainly the complete local ownership (in 1978) through P. T. Timah, and also the use of local petroleum products. In gravel pumping profitability was surprisingly low, given the highly intra-marginal nature of almost all other South-East Asian tin production, but RV was high mainly as a result of the intensive use of labour and the use of local power. The high RV figure for Indonesian tin is confirmed by an estimate made by P. T. Timah that the corporation's total imports in 1977 were \$22.9 million, roughly 9 per cent of total export revenue.

The generally high RV figures in South-East Asia can be compared with the estimates for Bolivia for 1974 by the Harvard team headed by Gillis (1978, ch. 3),[50] and with RV estimates produced for the World Bank by Ayub and Hashimoto (1985, ch. 3) based on data for 1980–1.

Gillis was interested in the prospective multiplier effects of expansion in Bolivian mine production, and of increases in price,[51] and the team estimated an RV of 45 per cent for an output increase at constant prices, and 52 per cent for a price increase alone. These estimates (which are for all non-fuel minerals and include private sector production, but are dominated by tin) are surprisingly low, given that most Bolivian production is controlled by Comibol. However, Gillis' definition of RV is different from that used here and from that of most other commodity studies in that its 'first round' refers to expenditure on locally-produced goods and (non-factor) services rather than income accruing to domestic income recipients. In particular Gillis deducts from RV the imports purchased by export sector workers and Bolivian owners of capital, and savings made by them, and also imports purchased by the government out of export tax revenue. Adding these items back into the RV estimates gives figures of 66 per cent for the output change. Though this figure makes only a very slight allowance for foreign ownership of capital in private sector mining in Bolivia, it gives nearly 100 per cent for the price change, the benefits of which accrue almost entirely to the government, local workers and local owners of capital. The fact that the Gillis estimates are for marginal propensities whereas the figures here are averages for the whole of a sector is not in practice an important difference, since almost all of the Bolivian marginal estimates assume the average and the marginal values are the same and are derived from statistics for averages.

Gillis's RV figures can be compared to those of Ayub and Hashimoto (1985, ch. 3), who argue that changes in energy costs and greater use of domestic inputs justify an updating of the Gillis estimates. Working with a sample covering at least 90 per cent of Bolivian tin production, they estimated that 53 per cent of gross revenue in 1980 and 1981 from increased tin production would be spent initially on domestic goods and services. With regard to a rise in tin export revenue resulting from a rise in the tin price, the figure is estimated to rise to 65 per cent in 1980 and 70 per cent in 1981.

RV is usually presented as a method of measuring the net foreign exchange contribution, with an implication that it indicates the real income gain generated by (marginal) mineral export activities.[52] However, the net foreign exchange contribution, which can be inferred directly from the RV figures set out above, is relevant principally where the country in question suffers from a chronic balance of payments deficit which it cannot cure by traditional methods. All three South-East Asian producers in the late 1970s had reasonably strong balance of payments positions. Bolivia's position was much less comfortable, with high external debt service ratio and a large current account deficit.

To assess the real income contribution of an export activity involves evaluating the real resource costs of the components of RV. In Gillis' (1978, pp. 68–9) account of Bolivia, the RV figure for changes in output was assumed equal to the real income gain by several simple devices. First, it was assumed that the labour used had no opportunity cost, and second, it increased the assumed cost of petroleum to take into account the fact that it was sold domestically at below world market price. Inputs were mostly imported and those were taken at their foreign exchange cost. This, as noted, gives a figure of roughly 66 per cent when adjusted to compare with the South-East Asian estimates; though for the Bolivian mineral export sector as a whole (see earlier) the real income gain was equivalent to a third of net export revenue when ramifications elsewhere in the economy are considered. Estimates of real income gains based on data from project appraisals in tin-mining by Thoburn (1981a, ch. 6) are included in Table 5.3. All inputs (including labour) have been revalued at efficiency prices (using the Squire–van der Tak (1975) project appraisal methodology developed for the World Bank); capital costs (also valued at efficiency prices) are charged at an assumed 10 per cent rate of interest. These costs are deducted from the value of output at the 1978 average tin price and the difference is expressed as a percentage of output value, yielding a *net economic gain* (NEG) coefficient. These coefficients, since they are drawn from individual tin mining projects, may not be as typical of their sectors as the RV estimates.[53]

Generally the costs in financial terms were higher than those in economic efficiency prices (except for Indonesian petroleum sold in the 1970s to tin-mining at below the world price). For Malaysian dredging, about 8 percentage points should be deducted from the gain coefficient to take account of the outflow of dividends to abroad (Thoburn, 1981a, p. 95), and for gravel pumping note that the small mine was still the more common sort to Malaysia. The estimates illustrate the generally high economic efficiency of tin-mining in South-East Asia in the 1970s, especially in Malaysia.

The RV and NEG coefficients include local payments of export duty/royalty and tax. It is worth asking whether the high rates of export duty charged in three of the main producers (Bolivia, Thailand and Malaysia) increased local gains by shifting the incidence of the tax on to foreign consumers and increasing revenue in the face of inelastic demand. A study by Lam (1978) of tax incidence in Malaysian tin argued against this proposition, on the grounds that individual Malaysian miners faced a virtually infinitely elastic demand and that they had a very low supply elasticity. In fact, the imposition of a heavy export royalty by Bolivia from 1965[54] may well have curbed production by depriving Comibol of replacement investment, and marginal (gravel pump) mines in South-East Asia could have been adversely affected by heavy taxation on output (Thoburn, 1981a, pp. 127–8). Moreover, estimates by Bird (1978), suggested grounds for believing that the estimates of tin supply elasticity in the studies cited by Lam were too low. Thus heavy export taxation may have reduced supply and may have been shifted in part to consumers. The very high prices of the late 1970s provide some support for this possibility, though the tax level and structure were by no means the only adverse influence on supply during that period. Regarding demand elasticity, it is doubtful that a large production cutback by a major producer would have left price unchanged. The effects of profits tax in South-East Asia (it was not used in Bolivia) also could have reduced supply, but its effect would have been much less.

Backward Linkages

The large mineral rents generated by tin in South-East Asia in the 1970s made tax and ownership policy an important part of securing development gains. Scope also existed for backward linkage creation, though much of this had come about naturally through market forces. Purchases of power in alluvial tin-mining usually accounted for between nearly 5 per cent and over 15 per cent of gross output value. Gravel pumps and offshore dredges normally use diesel and onshore dredges use electricity. In Indonesia electricity and diesel, in roughly equal proportions, accounted for 60 per cent of total intermediate purchases by the tin-mining sector in 1971 (*Tabel Input–Output, Indonesia 1971*), and this understates the importance of diesel which was sold in Indonesia in the 1970s at below the world price. In Malaysia fuels constituted 41 per cent of dredging's purchases of materials and 51 per cent of those of gravel pumping in 1974 (*Census of Mining Industries, Peninsular Malaysia 1974*). Indonesia and Malaysia are net exporters of petroleum products. In 1971 mining purchases constituted about 25 per cent of the output of the Malaysian petroleum products industry. The existence of this large consumer may well have influenced the Malaysian government in its decision to give 'pioneer' (tax exempting) status to Esso and Shell to establish local refineries in the 1960s, and the available evidence suggests that petroleum refining in Malaysia was not a 'bad' linkage.[55] Electricity generation on a large scale dates from 1928 in Malaysia, and in the late 1930s tin-mining bought over 30 per cent of the units generated in the two main west coast states of Perak and Selangor. In Indonesia the mining islands have their own electricity generating facilities though, since the islands are far from the country's main centres of population, the facilities do not yield externalities to other industrial users. In Thailand, mines in the main tin area in the south purchased electricity on a large scale, but the development of the industry was not investigated.

Railway development in Malaysia also depended heavily on the tin industry, in the sense that the railways were built over the 1884–1910 period largely out of government revenue, over a third of which came from the export duty on tin, a good example of 'fiscal linkage' (Hirschman, 1977, pp. 72–7). In colonial Malaya, the early stages of railway development had taken the form of lines from the tin areas to the ports, and the national network was formed largely by joining these up. In Thailand although railways had been built by the government since the late nineteenth century, and a line had linked the South (the main mining area) with the central region since before the second world war, one cannot link railway development to tin in a similar fashion. Tin never occupied the dominant position as a source of revenue that it did in Malaya in the early years of the industry's rapid development, and tin ore was shipped directly out to Malaya for smelting. In Indonesia tin was shipped abroad directly from the mining islands.

In Malaysia probably the most interesting linkage, though it owes little to government action, is the growth of a local light engineering industry to meet the needs of the tin industry both for spare parts and for capital equipment. Engineering parts are the most important intermediate product purchased by the tin industry after fuel, and local engineering firms date to before the first world war. With regard to capital equipment, in the 1970s about a quarter of the value of an inland tin dredge and about 40 per cent of the capital cost of a gravel pump operation consisted of locally-purchased engineering and construction items. In Indonesia, extensive facilities existed for the manufacture and servicing of mining equipment at P. T. Timah's workshops on Bangka, which were capable of doing dredge rehabilitation and which dated from Dutch colonial times. They did not extend to major new construction, however, and two offshore dredges commissioned in the 1960s and 70s were built overseas (in the UK and Japan) and towed to site.

This is also true of offshore dredging in Thailand, though in Malaysia dredges were normally fabricated on site. In Thailand local workshops in the Phuket area were producing a range of spare parts (aided by a 15 per cent import duty). In Malaysia and Thailand most of the engineering industry was in the hands of locally-owned (and mainly ethnic Chinese) firms, though dredge design was normally done overseas. Malaysia dredge erection contractors have also carried out work in Thailand and Indonesia.

In alluvial mining backward linkage generation is qualitatively similar between the various alternative techniques, though (as already noted) dredging tends to use electricity as fuel and gravel pumping uses petroleum products. Gravel pumping tends to be the more materials-intensive. In Thailand this led to a larger outflow of export revenue than in inland dredging, since Thailand is not a petroleum producer, but it also led to a larger demand by the gravel pumping sector in all three countries for engineering parts. The demand for engineering products was also greatly enhanced in Thailand by the growth of suction boat mining, the boats being of local construction. A detailed case study of engineering in Malaysia (Thoburn, 1973b) showed little evidence of its being a 'bad' linkage. Indeed, it proved a useful vehicle for the development of small and labour-intensive businesses. Further scope would have existed for local manufacture of mining supplies, in all three South-East Asian producers, but careful project appraisal would have been necessary where they were to be made as part of an import substitution policy behind tariff barriers.

In Malaysia only a quarter of the mining industry's purchases of intermediate products (which were overwhelmingly for tin-mining) was imported directly.[56] In Bolivia, in contrast, Gillis' team (1978, p. 76) estimated that 85 per cent of the material purchases of non-fuel mining were imported.

Finally, most of the backward linkages generated by tin, at least in South-East Asia, are such as to outlive the eventual exhaustion of production. The provision of power relies increasingly for its market on other industrial development, and some engineering firms can find a wide range of customers outside of the mining sector. The evidence in Malaysia is clear, however, that most engineering firms supplying the mining industries owed their establishment and early growth to mining even though by the 1970s they had a wider range of customers (Thoburn, 1973b). This ability to find other customers was sorely needed in the 1980s and 1990s as the tin price collapsed.

Forward Linkages

Forward linkages involve moving from a consideration of RV to looking at the producer countries' share of Final Consumer Value (FCV).

The local development of tin smelting has already been traced in detail.

All South-East Asian tin output by the 1970s was being smelted in the region. Smelting adds only about 1–5 per cent of output value in alluvial mining, but it saves 25 per cent weight in shipping. Smelting improves the countries' bargaining position relative to purchasers in the sense that there are many more buyers for tin metal than there were for concentrates. The Malaysian smelters at least, have been extremely profitable historically. It was estimated in 1974 that the Straits Trading Company made a M$12 million profit on a probable net turnover of M$18 million (M$740 million gross turnover less payment of M$722 million for supplies of unsmelted tin). High past profits have been put into investments, including some mining companies. Straits Trading derived almost half its profits from investment (Straits Trading Company, AR, 1977). In 1969, Datuk Keramat owned investments of M$31 million compared with its own total market value of M$75 million (Thoburn, 1977b, pp. 123–7). Smelting charges in Thailand were geared to those of the Malaysian smelters, but data on Thaisarco's profitability for the 1970s are not available, nor on that of Peltim smelter in Indonesia. However the processes used (after some initial experimentation in Indonesia, already discussed) are basically similar to those in Malaysia, and there is no reason to believe that the Malaysian smelters' high private profitability greatly overstates their social profitability since their output is exported on a free market and their inputs are not subsidised. The problems faced by Bolivia of establishing domestic smelting capacity for her complex, hard-rock ores have already been discussed and there may well have been some trade off between improving the country's bargaining position against foreign buyers and the high real costs of domestic smelting.

Forward integration into the manufacture of tinplate and tin cans is the other main potential linkage, and there is a local demand in South-East Asia not only by consumers, but from export industries selling canned fruit such as pineapples. The problem here is that tinplating is more closely allied with the local establishment of steel-making than with tin-mining (since tin metal is a tiny part of total cost). Tinplate for can manufacture can be imported, so neither activity is greatly aided in cost terms by the existence of local tin-mining. It contrasts to the backward linkages of electricity generation, where lack of transportability gives natural protection to domestic capacity, and engineering where local contact with customers is extremely important. This is relevant both to the reasons why the mining sector may have prefered to buy locally rather than import, and to why the backward linkage industries did not set up initially as export industries in their own right. A study for UNIDO by David Lim (1980) was extremely pessimistic about tinplate manufacture, and gave instructive data about the tinplate industry in Thailand, the only major tin producer at the time to have one. The Thai industry consisted of one firm, which was a joint venture between local interests and a Japanese steel manufacturer, and ran at under a third capacity while more than half of Thai tinplate consumption was imported. The main problem was that much of Thai demand was for lower quality ('secondary') tinplate whereas the local company produced 'prime' tinplate. However, nor could it export to neighbouring Malaysia and Indonesia because of high costs and their preference for 'secondary' tinplate, and Lim did not feel it had prospects of exporting to industrial countries. In any case, even if all the tinplate consumed in Thailand (e.g for local pineapple canning) had been produced locally, it would have required only about 3 per cent of the Thai production of tin metal. In Indonesia domestic demand was smaller still and below the efficient minimum production for an electrolytic plant. Tinplate has a high income elasticity of demand (Lim, 1980, p. 210) so LDC tin producers needed to wait until their domestic markets grew. However, Brazil's tinplating industry appeared to work at near capacity, and was expanding in the 1970s (*TI*, January 1980). Malaysia's demand for tinplate grew rapidly too, apparently (at 10–20 per cent per annum in 1970s),[57] and at the end of the 1970s a major tinplate plant (Perstima) was to be established as a joint venture between the government-owned food corporation, FIMA, and Japanese steel interests. There was a minority holding (35 per cent) taken up by MMC, the Johore State Economic Development Corporation and some pineapple canners (*TI*, November 1979). Indonesia was planning to introduce tinplating in a steel complex in Java, but Lim cited a 1972 feasibility study by P. T. Timah which reported strongly against local tinplating. Bolivia in the 1970s had no tinplate facilities.

Besides tinplate, solder appeared in the 1970s to have prospects, especially in Malaysia where there is a flourishing electronics industry. Tin, though constituting 60 per cent of the weight of the solder, is only a minute component in the total cost of the final electronics product again, the proportion of local metal output which would be taken would be very small.

In Malaysia, where the first domestic use of tin was recorded in 1977, it was used mainly for pewter, and almost all of the rest was for solder; the use was equivalent to less than 0.5 per cent of Malaysian tin-in-concentrates production. By 1983, with the first Malaysian production of tinplate, domestic metal use doubled, and was equivalent to 1.9 per cent; domestic use doubled again within a year, bringing the proportion of domestic use in relation to output to 3.8 per cent. By 1990 domestic use of tin had almost doubled again, and this, combined with falling tin output, raised the proportion of tin production used locally to 9 per cent. In 1992 half of the local use was for producing solder, and less than a fifth was for tinplating. Domestic production of pewter, mainly for tourist souvenirs, remained a scarcely less important use than tinplate (*MT*, April 1993). In Thailand too, local consumption of tin metal has risen considerably since the 1970s. From 1981 to 1986 domestic consumption of tin metal doubled, and increased as a proportion of mine production of tin from 2.5 per cent to 8.9 per cent. By 1990 domestic tin consumption had almost doubled again, and this, combined with falling mine output, raised the proportion to 19.1 per cent (figures supplied by DMR, Bangkok, 1992). In Indonesia in 1990, domestic consumption of tin metal was equivalent to 3.9 per cent of mine production of tin-in-concentrates. In Brazil for 1990 the figure was 22.3 per cent, but for Bolivia only 1.2 per cent. China was estimated to consume an amount of tin which was equivalent to 38.6 per cent of its mine production, while the former USSR at that time consumed almost double the amount of tin which it produced (ITC, *ITS*, July 1992).

Final Demand Linkage

In South-East Asia, data on local market creation by tin mining workers' consumption expenditures are available only for Malaysia, where it was estimated that roughly 70 per cent of their expenditure in the 1970s was on locally-produced goods (Thoburn, 1977b, p. 179). This high local proportion also could have been true for Thailand given the ethnic similarity of the workforce to that of Malaysia, though obviously consumption would be influenced too by the pattern of import protection. In all three Asian producers the absence of large numbers of really highly-paid workers found in some mining industries elsewhere again predisposes one to expect a low import content in expenditure. In Bolivia, the Harvard team estimated that 75 per cent of mine workers' expenditure was local (Gillis *et al.*, 1978, p. 76).[58] This figure is confirmed almost exactly (73.9 per cent local expenditure) by Ayub and Hashimoto (1985, p. 25) on what they regard as more reliable data.

Urbanisation is also an important feature of local market creation. This is not to argue that urbanisation is an unmixed blessing, as Lipton's (1977) work on urban bias has shown; simply that mining in remote areas is likely to generate less development than is mining geographically linked to the rest of the country. In Malaysia urban growth was associated with tin mining. Most of the present day major cities, such as Kuala Lumpur, Ipoh and Seremban, were originally mining towns. In Thailand and Indonesia, in contrast, tin-mining areas were far from the other main centres of population. In Bolivia the mountainous region has both the bulk of the country's population and almost all its mining, but this area is far removed from the agricultural areas of the country and difficult to reach from them, so to that extent it has not led to the economic integration of the country. The transport system has been geared to shipping ore out via Chile, whereas in Malaysia the rail network built to transport tin has formed the nucleus of

TABLE 5.4   Direct employment creation in South-East Asian tin projects

		Workers per 10 tonnes annual output	Workers per US$1 million of investment
Malaysia	Secondhand inland dredge	2.02	12.08
	Large gravel pump mine	5.24	87.60
	Small gravel pump mine	7.35	78.42
Thailand	Secondhand inland dredge	12.5	63.98
	Small gravel pump mine	7.35	70.52
Indonesia	Offshore dredge	2.17	8.26
	Small gravel pump mine	5.74	110.53

Source: Thoburn (1981a, p. 115)

the present-day transport system. Its role in the past in facilitating the spread of rubber cultivation and other economic activity in Malaysia is well documented (Thoburn, 1977b, ch. 6).

Employment Creation and Unpriced Externalities

Table 5.4 gives an indication of the employment which would have been created by new tin mining projects in the various sectors in South-East Asia in the late 1970s, and Table 5.5 indicates the employment intensity of tin-mining sectors in the main producing countries

Both in the new investment projects in Table 5.4, and the sectors as a whole in Table 5.5, gravel pumping in South-East Asia had a much higher employment intensity generally than dredging, especially dredging offshore. Both Comibol in Bolivia and Timah in Indonesia ran their operations at a somewhat higher labour intensity than commercial profitability would suggest. With regard to investment, the very high employment creating potential of gravel pumping (Table 5.5) is striking, though one should remember that its capital equipment would have been shorter lived than that of a dredging project. The employment-creating effect of suction boat mining in Thailand in the 1970s (though not shown in the ITC statistics reproduced in

TABLE 5.5   Sectoral employment intensities in tin mining, 1978

	Workers per 10 tonnes annual output		Workers per 10 tonnes annual output
Malaysia		**Bolivia**	
Dredging	4.62	Underground	6.58
Gravel pumping	7.86		
Thailand		**Australia**	
Offshore dredging	1.40	Underground cast	1.27
Onshore dredging	11.23	Open cast	NA
Suction boats	NA		
Gravel pumping	18.71	**United Kingdom**	
		Underground	3.48
Indonesia			
Offshore dredging	2.39		
Onshore dredging	7.26		
Gravel pumping	8.35		

Sources and Notes
1. Based on figures from ITC, *Monthly Statistical Bulletin*, May 1979

Table 5.5) greatly exceeded that of offshore dredging, both in output terms and investment terms, though again project life was shorter (Thoburn, 1981a, p. 136).

Of the unpriced externalities generated by tin, labour training is probably the most important. About 40 per cent of the workforce in a Malaysian dredging operation was skilled or semi-skilled. In gravel pumping the figure was nearly 30 per cent (Thoburn, 1977b, p. 213), though the higher overall labour intensity of gravel pumping would mean that the overall use of skilled workers per unit of output and per unit of investment is greater by the gravel pumping sector. In Thailand and Indonesia there is no reason to suppose a radically different skill mix in each of the sectors from that of Malaysia. Some evidence collected on new tin investment in the 1970s by Thoburn (1981a, ch. 6) suggests that the differential between mining wages and shadow wages in the economy may have been greater in Thailand than Malaysia. Casual empiricism suggests that this would also have been true in Indonesia (shadow wage data for Indonesia were very rough and ready at that time).

Clearly in the short run the increased use of skilled workers as a result of an output expansion in tin would be a cost to the economy, but in the longer run a higher demand for skills tends to be met by an increased supply and an overall improvement in the quality of the workforce. Moreover, the skills used in tin-mining are generally usable by other sectors if the workers wish to move; drivers, electrical chargemen, skilled workshop workers, for example, are prominent in the skilled mining workforce. In the

1980s and 1990s this has allowed skilled workers in Malaysia and Thailand to move easily into other jobs in those countries' rapidly-expanding economies as large numbers of tin mines have closed. Such redeployment has been more difficult in Indonesia because of the geographical isolation of the tin islands, but redeployment has been aided in Thailand by the growth of the tourist industry and other activities in the southern provinces.

The skilled nature of the mining workforce is reflected in relatively high rates of pay in Malaysia in the sense that gravel pumping workers in the 1970s were receiving slightly higher wages than manufacturing workers on average, and dredging workers' wages were 50 per cent higher. However, the employment intensity in tin-mining in Malaysia in the 1970s was roughly equal in dredging and manufacturing, while gravel pumping's employment intensity was higher, so moderately high wages were not accompanied by a particularly low level of labour use.[59]

In Malaysia there have been no 'Europeans' (westerners) in the artisan, junior supervisory or clerical positions since the very early days of the industry and the use of Thai nationals to replace expatriates dates from the second world war. Even dredgemasters in Malaysia and Thailand were, by the 1970s, often local, and in Indonesia local people in P. T. Timah have carried out all of the functions since nationalisation, as in Bolivia.

Data on the skill mix in Bolivian tin-mining are not readily available, but the low living standards of workers in Comibol and other Bolivian mining companies have been noted by a variety of commentators. A British trades union delegation (National Union of Mineworkers, 1979) remarked on the 'appalling' wages of Bolivian mine workers. They noted that miners' life expectancy was on average only thirty-five years, while silicosis and tuberculosis were endemic and mine safety precautions were almost non-existent.[60] Thus the low accounting profitability of Comibol, the high costs caused by the nature of the tin deposits, and the lack of reinvestment in the post-war period have resulted in workers enjoying lower real incomes than most tin mine workers in Asia.

OVERVIEW: THE 1960S AND 1970S

The tin industry in the 1960s and 1970s experienced a rising real price trend. At the time, this was thought to be a reflection of the fact that tin reserves were low in relation to annual consumption, as compared to other major metals. Over the years, tin had lost market share, particularly to aluminium whose real price had been falling while tin's had been constant. Malaysia, the world's largest producer, after a peak output in 1972, showed a declining output trend despite the rising tin price. Though this fall in Malaysian production was attributed by many miners to excessive taxation, it was more a reflection of the fact that existing mining land had been already heavily worked, and new leases were hard to obtain. These difficulties were not simply geological; they were also rooted in conflicts of interest between central government and individual Malaysian states. Although some tin mining was taking place in Rondonia in Brazil, there was virtually no hint of the expansion in output which was to come from Brazil in the 1980s.

The high tin prices meant that intra-marginal producers were able to earn very large mineral rents, in the sense of profits over and above the supply price of their investment. Although there were of course variations in cost between particular mines in each producing country, a distinction between the generally low-cost mines of South-East Asia and the higher cost (underground) mines of Bolivia seemed clear. Bolivia's high costs may have been exaggerated by an overvalued exchange rate, but they also sprang from the inaccessible nature of Bolivian tin deposits. Simply using depreciation of Bolivia's real exchange rate to lower the price of Bolivian tin exports probably would have purchased increased competitiveness at the expense of lowering Bolivian mineworkers' already low real incomes. In any case, real prices tended to cover Bolivian costs. Given the fairly large disparity between the supply price of different producers, if ever tin demand had fallen to a level which did not require the production of Bolivia, or if there had been new, low-cost sources of supply, prices could have fallen substantially. Both these events occured in the 1980s, but were not foreseen.

The high mineral rents led to considerable interest in the two main 'free-market' tin producers, Malaysia and Thailand, in how to use taxation and equity participation by government as means of securing for the host economy a larger share of the gains *vis-à-vis* foreign investors. In Malaysia, control over the main foreign producer, the London Tin Corporation, which also still had interests in Thailand and Nigeria, was acquired by a Malaysian government agency in a stock-exchange takeover. This led to the formation of the Malaysia Mining Corporation. The Malaysian takeover was part of the New Economic Policy, which was designed to restructure ownership in the economy in favour of indigenous Malays. It was politically pressing to achieve this because of the serious racial riots of 1969, where Malays, feeling disadvantaged, had attacked Chinese in Kuala Lumpur and other cities following an acrimonious election campaign. Foreign control of the primary export sector was an important target for restructuring, and guidelines were also laid down about Malay employment. In Thailand similar restructuring followed the temporary overthrow of military government. Thailand and Malaysia also had in place taxation of profits, and a sliding scale export duty, the latter dating back to much earlier times. Tin-mining in Nigeria and Zaire during this period was in decline, speeded in Nigeria's case by the exchange rate effects of expanding oil exports.

Breaking with the rest of the book's chronological treatment of the various themes of interest in the tin industry, this chapter has presented a single case study of the development effects of tin-mining in South-East Asia, drawn mainly from earlier work, updated where possible, and on Bolivia from secondary sources. Various measures are used of the gains from tin-exporting, including coefficients of Retained Value and Net Economic Gain. There is also discussion of the linkage effects of tin and of its wider development effects such as labour training and employment creation, as well as of the non-marginal effects of Bolivian tin exporting.

In the late 1960s, before the restructuring of ownership in Malaysian tin dredging, RV was over 70 per cent,[61] and 77 per cent by 1978. Dredging had a very high proportion of profits in gross output value, with a small wage bill and a small but not insignificant local expenditure on materials. The channeling of income back into the domestic economy was therefore a result of tax payments. Local tax payments are a post-war phenomenon, although export duty dates back to the earliest colonial times. Local purchases of dredging shares also raised RV in the post-war period. Although labour use per tonne of tin almost halved between 1938 and 1972 (Thoburn, 1977a, p. 41), the share of wage payments in output value increased, reflecting rising real wages. RV in gravel pumping was higher (about 85 per cent through the 1970s), as would be expected from a locally-owned sector. The higher RV figure also reflected larger local wage payments and materials purchases. The tin industry generated linkages in the form of power; first coal (in the early years of the industry), then electricity and petroleum products. Engineering, both for spare parts and capital equipment, was also an important linkage. RV in virtually all other South-East Asian tin mining in the late 1970s, with the exception of Thai offshore dredging, was about 80 per cent or more.

However, other minerals produced in the Third World have high RVs.[62] In studies by Mikesell and his associates (1971) of foreign investment in minerals and petroleum, in nine commodity and country case studies RV was typically 65–70 per cent. Among Pearson and Cownie's (1974) eleven African case studies (which had average RVs of 76 per cent, though these included some agricultural exports), foreign investments in Nigerian petroleum and Zambian and Zairean copper all had RVs over 70 per cent. High RVs in minerals generally reflect heavy local taxation. For example, taxation was mainly responsible for raising RV in Chilean copper from 35 per cent in the 1930s to 94 per cent in 1953, but it was at the expense of discouraging foreign investment. Various reservations about RV as a measure of development gains have already been expressed in this chapter and need not be repeated. While very low RV is likely to indicate that gains are being lost, higher RV may have been achieved, as in Chile, at an unacceptable cost. In

Bolivia moderately high RV (about 66 per cent in the 1970s) was accompanied by a lack of international competitiveness and heavy overstaffing. In most South-East Asian tin-mining activities in the 1970s, high RVs were accompanied by high Net Economic Gain coefficients too, except for onshore dredging in Thailand and gravel pumping in Indonesia, both of which suffered from some technical inefficiency.

Some other mineral exports have generated local linkages, as is the case in the Zambian copper industry, but tin is unusual in the very extensive nature of its stimulus to local capital goods production and in its high labour intensity. Tin's high labour intensity, compared to import substituting manufacturing for instance, has neither been accompanied by the exceptionally low real wages of the Bolivian tin mineworkers, nor the very high wages (normally paid to a very small labour force, however) of the copper industry, which can result in a highly dualistic wage structure and lopsided development for the economy.

The difference in development effects of tin as between the South-East Asian case and the Bolivian situation can to a large extent be explained in terms of the almost total dissimilarity of underground lode mining from surface alluvial mining. Also, of course, the long-standing discrimination against Bolivian Indians, who constitute most of the mining labour force, and the lack of alternative employment opportunities (except now in coca growing), have been forces in keeping mine wages low. In Malaysia, tin-mining workers have had easy alternative employment opportunites in both the nearby urban and rural economies. Historically this has meant higher wages in Malaysia than in the remote tin islands of Indonesia too. Nigerian tin mining has been chronicled in a well-known study by Bill Freund (1981), and Freund's argument is that tin's impact on Nigeria has been largely negative. This contrast to South-East Asia and cannot be explained away by differences in the type of mining, since the techniques are similar to Malaysia's in many respects, but by the social, political and economic circumstances in which tin was introduced. Chapter 7 returns to this issue.

Development effects since the 1970s obviously will have been affected adversely by the loss of mineral rents in South-East Asian as the tin price collapsed, as Chapter 6 will chronicle, and by the great falls in employment in the gravel pumping sector. The growth of tin-using industries in tin-producing countries has been the main new development since the 1970s.

NOTES

1. This section is a shortened and revised version of Thoburn (1981a, ch. 5, pp. 89–113). The preceding sections also derive most of their material from chs 4 and 5 of this source, which includes information from interviews I conducted in Europe and South-East Asia in 1978–9. My earlier book contains tables which give more

detail on the payments structure of tin mining than those presented here, and there is more information on individual companies, especially in Malaysia and Thailand.

2. The original source of Table 5.1 (Williamson, 1984) gives estimated figures for 1983 and 1984. In updating Williamson's 1983
   and 1984 figures using ITC (1986) statistics, some differences
   emerge between his table and the ITC statistics for earlier years.
   Williamson's original figures have been retained for earlier years; in
   any case, the ITC revised some of its figures from year to year. It is
   assumed that Williamson's 'production' statistics refer to tin metal
   rather than tin-in-concentrates production. Note that the ITC inter-
   ventions in the market in the run-up to the ITA's collapse were dis-
   guised by various means discussed in Chapter 6.
3. Calculated from Raphael Zorn (1976).
4. See Lanning and Mueller (1979, p. 278) for further discussion.
5. See Lanning and Mueller (1979, chs 12 and 19 especially), who are
   especially useful on Anglo's methods of securing effective control
   at minimum cost via minority shareholdings. See also Pallister,
   Stewart and Lepper (1988) for a more general study of
   Anglo–American, which concentrates on the group's operations
   within South Africa.
6. Amalgamated Tin Mines of Nigeria (which produced about 45 per
   cent of Nigerian tin output in 1973 and in which LTC owned 30
   per cent) accounted for 2.4 per cent of LTC assets, and its Thai
   companies 7.9 per cent.
7. The total number of mines in 1976 was given as 538 in *Mineral
   Statistics of Thailand*, but only 4 suction boats were listed. Since
   there were probably several thousand suction boats at work on a
   quasi-illegal basis, the importance of small-scale operations in total
   numbers is thus understated.
8. According to the Malaysian Department of Mines (*BSMI*, 1975)
   'European' producers accounted for 54.9 per cent of open cast
   output and 82.5 per cent of underground mining output, but these
   two methods produced only 3.9 per cent and 2.9 per cent, respectively, of Malaysian tin output. In dredging and gravel pumping
   the 'European' shares of production were 92.7 per cent and 11.9
   per cent, respectively. The basis for tin definition of production as
   'European' or 'Asian' is not given in the statistics, but it probably
   follows other official usage in referring to whether over 50 per cent
   equity ownership is owned by Europeans or Asians.
9. Thoburn (1977a, p. 43) shows that in 1974 the rate of return on a
   dredging investment was higher than a gravel pump operation at
   the then price of about M$1000 a picul (M$16.54/kg), but it would
   have been lower than the gravel pumping rate of return at a price
   25 per cent higher. At a 15 per cent discount rate, the net present
   values also switched in favour of gravel pumping. M$16.54/kg is
   substantially higher than the tin price of the early 1990s. Since
   1974, fuel costs will have fallen in real terms, but other costs will
   have risen with inflation.
10. The All Malaya Chinese Mining Association in a submission to the
    Malaysian Ministry of Finance in 1978 estimated that 80 per cent
    of its members were working on tailings or previously dredged land.

11. Output in 1972 was 76,830 tonnes compared with 57,767 tons in 1950. There was, however, sharp export control under the ITA 1957–60 and lesser control in 1968–9, 1973 and 1975.

12. The TEMCO dredge, delivered in 1971, had a capacity of about 240,000 cubic metres per month. The Aokam 3 sea dredge was almost certainly the only other new dredge purchased since the war. Its capacity was 420,000 cubic yards a month, (Rachan, Pow and Champion, 1969), about half the capacity of the new Malaysian dredges of the 1970s.

13. In 1974 offshore dredges produced 22 per cent and inland dredges 10 per cent of Thai tin output. In 1976 these proportions had fallen to 7 per cent and 9 per cent, while suction boats (shown in Department of Mineral Resources statistics for the first time that year) produced nearly 23 per cent of total tin output.

14. Kamunting ceased its operations in Thailand in 1975.

15. Fairmont State was the new name for Faber Merlin, which also had had links with the Pahang Consolidated Corporation, the largest underground mine in Malaysia. In 1976 Faber Merlin owned 25 per cent of PCCL (Raphael Zorn, 1976).

16. Fox (1974, p. 78) notes that in the mid–1960s 'New management and fresh capital were pumped into South Crofty from the Siamese Tin Syndicate (now renamed the St Piran Mining Co.)'.

17. The 1969 Annual Report of P. N. Timah listed unsuccessful negotiations with Charter Consolidated and a consortium of Rio Tinto Zinc and Bethlehem Steel from the US. Bethlehem tried in the late 1970s to establish an inland and an offshore tin dredging venture in Thailand, presumably to secure a tied source of tin.

18. Information from interviews with P. T. Timah and the three companies, April 1979.

19. See Gillis (1978, pp. 134–5) and Gillis and Beals (1980, ch. 4) for discussions of Indonesian mining contracts. Only one 'first generation' contract was signed. Sixteen 'second generation' contracts were signed, 1968–76 (US Embassy Indonesia, 1977).

20. Only one such had been signed by the late 1970s (with Rio Tinto Zinc on a copper project) and it included a 60 per cent windfall profits tax when the return on capital rose above 15 per cent over a three-year period, together with the obligation to offer 51 per cent of tin equity (instead of the usual 25 per cent) within ten years of starting production. Contracts thereafter were expected to have still tougher provision, including stipulations about the use of local contractors (*FEER*, 1 February 1980).

21. Details from an interview in Brussels, June 1978.

22. Ownership details for all the major Australian tin mines are given in McKern (1976, pp. 238–9). Company output figures in the 1970s were given monthly in *Tin International*.

23. More details of the Malaysian takeover of the London Tin Corporation are given in Thoburn (1981a, pp. 82–4)

24. See Charles Raw's (1978) 'exposé' of Slater Walker, though Raw unfortunately has little to say about its Asian activities.

25. Further details of the takeover and subsequent events are given in *FEER*, 1 April 1977, 7 September 1979 and 30 November 1979.

26. See text of speech on Mineral Development Policy by the Minister

of Industry, 22–3 March 1978, translated from Thai by the Department of Mineral Resources, Bangkok.

27. Speech by chief minister of Selangor, July 1978, printed in *Tin*, November 1978.

28. The smelter had held 10 per cent of the shares of Malayan Tin Dredging, which merged with MMC in that year, and in return Datuk Keramat was given 3.8 per cent of MMC (Elliott, 1989, p. 47).

29. The 1971–7 figures, on which this statement is based, are from a study by D. J. Fox quoted in detail in *TI*, December 1978. More detailed figures for 1973–4 are given in Gillis *et al.* (1978, pp. 329–31) and for 1978 in Engel and Allen (1979, p. 127).

30. This is less than the amount spent by a number of mining companies in Indonesia on single claims (Gillis *et al.*, 1978, p. 50).

31. According to Engel and Allen (1979, p. 153) foreign ownership in private sector tin production in Bolivia was limited to 49 per cent.

32. Information on Timah's activities in the 1970s, unless otherwise stated, comes from interviews in Jakarta in April 1979.

33. Prior to 1976, Timah paid 25 per cent of its after-tax profits to the government (as a contribution to the National Development Budget). In 1979 Timah's dividend payment was fixed at 20 per cent.

34. Based on data in P. N. Timah (1972) and P. T. Timah (1976).

35. Outputs from Department of Mines and Energy, Indonesia (1977) and costs from P. T. Timah, *Annual Financial Report*, 1977.

36. In 1973–4 Comibol produced 66 per cent of Bolivia's lead production, 88 per cent of zinc and 91 per cent of bismuth (Gillis *et al.*, 1978, p. 32).

37. Certainly some workers on contractors' mines I visited in April 1979 were Chinese, and signs on some mines were written in Chinese.

38. See Thoburn (1977b, p. 127) on the Straits Trading Company. In 1978 the large trading company Amalgamated Metal Corporation, which owned the other Malaysian smelter, Datuk Keramat, stated in its annual report that tin smelting generated 74 per cent of its profit while generating only 23 per cent of its turnover.

39. For Thailand no statistics are available to compare with Malaysia, but the Billiton company which owns the Thaisarco smelter claimed to make the bulk of its profits in the country from smelting rather than marketing.

40. Because a large part of Indonesian output comes from offshore dredges, total output is lower during the stormy season, and a larger smelting capacity is necessary to deal with this seasonal pattern than the simple figure for total annual output would suggest.

41. According to J.-F. Hennart (1986b, p. 247), the capital cost per ton for smelting capacity for lode concentrates is about three times more than for alluvial concentrates.

42. Whereas RV and the local share of FCV are in themselves exclusively marginal measures, the measures of some 'wider' development effects merge into the non-marginal. In particular, the income distributional effects of the mineral export activity clearly should be considered on a non-marginal basis.

43. Bolivia qualified as a 'mineral economy' on the 10 per cent share of national income criterion in 1970 (tin exports generated 10.6 per cent of GDP). By the mid-1970s the share had fallen slightly (8.5 per cent of GDP in 1975) (Baldwin, 1983, p. 43).

44. This rise would have to be in relation to non-traded goods (NTGs) prices, since the international terms of trade are held constant by assumption (i.e. Bolivia is assumed to be a 'small country' in the world economy), though the Harvard analysis does not make the role of NTGs explicit.

45. As the next subsection will show, however, the Gillis definition of RV is narrower than that generally used. Also, as discussed later, RV does not wholly represent a gain in real income to the host economy since the social opportunity cost of the resources used to generate the RV must be deducted.

46. These are based on cost breakdowns for 1978 set out in Thoburn (1981a, pp. 95–100). Data comes from cost figures supplied by the producer governments to the ITC. For Malaysia further very detailed splits of costs and payments were available from the Census of Mining Industries on a sectoral basis and from the Financial Survey of Limited Companies for tin mining as a whole; and these were used to estimate the local content of each item. For Thailand and Indonesia similarly detailed further breakdowns were not available and more use was made of general *a priori* information and rough estimates.

47. The details of RV and the input structure in earlier years are from Thoburn (1977b, pp. 95–111).

48. In 1978 this varied from 5 per cent to 12.5 per cent depending on the size of profits.

49. Note though that petroleum products have been refined locally in Thailand since the 1960s, and it is the crude which was imported (Ingram, 1971, p. 281).

50. The Harvard team, also headed by Gillis, which reported on the Indonesian mining sector in the late 1970s argued that the type of costs and payments breakdown required to arrive at a reliable estimate for retained value were not then available in Indonesia, either for foreign or domestic enterprises. Instead, Gillis presented an intuitive, very general argument that one would expect RV figures for Indonesia to be similar to those for Bolivia since they both had large state enterprises mining tin. No discussion in this context was given of the influence of the great differences between the tin deposits, mining techniques and costs in the two countries (Gillis and Beals, 1980, pp. 33–4).

51. Gillis' estimates of RV must necessarily be with respect to *changes* in export revenue since tin production is not marginal to the economy. In other words RV for the tin export sector as a whole in Bolivia is not a valid measure of real income gains. Instead, as the previous subsection has shown, the gains to Bolivia from the existence of tin-mining have to be estimated taking into account the effect of exchange rate changes on other sectors and on consumer welfare from imports.

52. For a discussion of the uses to which the RV measure can be put, see Brodsky and Sampson (1980). However, they modify it as real

income measure only with regard to social costs such as environmental damage, and do not discuss, for instance, the situation
where the social opportunity cost of labour employed in export
activity is below the market wage.

53. This is especially likely to be true for onshore dredging in
    Thailand, where project data in the Thoburn (1981a, ch. 6) study
    were difficult to obtain.

54. Although the export taxation of Bolivia is discussed here as if it
    consisted of a single export tax, it actually consisted of an export
    royalty and (from 1972) an export tax. The former made some
    allowance for presumptive costs and the latter for different grades
    of concentrate and (by very broad categories) different mining sec-
    tors. There was also some tax incentive for excess production. See
    Gillis *et al.* (1978, ch. 6) and Engel and Allen (1979, p. 156).

55. Thoburn (1977b, p. 126) cites an earlier (unpublished) study by C.
    B. Edwards to the effect that petroleum refining in Malaysia had a
    very low ratio of domestic resource cost to foreign exchange
    saving. Thus the 'inappropriateness' of its high capital intensity
    may have been offset by economies of scale and advanced technol-
    ogy.

56. This was according to the 1970 input–output table. In 1965 (using
    an inverse input–output matrix for the 1965 table calculated by Lo,
    1972, p. 57) imports represented only 3.5 per cent of output value
    in the table of cumulative primary factors.

57. In 1973 approximately 0.9 per cent of manufacturing value added
    in Peninsular Malaysia was generated by the manufacture of tin
    cans and metal boxes and 0.8 per cent by fruit canning (Malaysia,
    Annual Bulletin of Statistics, 1976).

58. Similar information was not available from the Harvard study of
    Indonesian mining in the 1970s. See Gillis and Beals (1980, p. 32).

59. However, gravel pumping in 1972 in Malaysia employed per unit
    of output only a quarter of the workers on rubber estates and only
    37 per cent of those on oilpalm estates (see Thoburn 1977b, ch. 9).

60. These poor conditions were also noted by the Harvard team sev-
    eral years earlier, and they felt things were worse in Comibol than
    in the private sector (Gillis *et al.*, 1978, pp. 38–9). Graphic
    descriptions of the poor conditions of life of Bolivian mineworkers
    are given in June Nash's two anthropological studies (Nash, 1979
    and 1992)

61. Some further information on Malaysia is from Thoburn (1973a
    and 1977b).

62. Comments on other primary commodities are drawn from surveys
    of empirical studies of trade and development in Thoburn (1977b,
    chs 10 and 11). See also Pearson and Cownie (1974) and Mikesell
    (1971) for useful sets of case studies. The available empirical case
    studies of trade and development studies mainly date from the
    1970s, when interest in this topic was at its height.

CHAPTER 6

The 1980s and 1990s
Price Collapse and Economic Restructuring

The tin industry in the 1980s and 1990s has been dominated by the collapse in prices following the financial failure of the International Tin Agreement in October 1985. Over the 1980s, tin experienced the largest price fall of any major primary commodity, and the tin price by the start of the 1990s was at its lowest level in real terms since the 1930s depression (refer back to Figure 1.1). The mineral rents, over which there was so much bargaining in the 1970s, were eroded for most producers, whose main concern became sheer survival. Underlying the collapse was not only the long-term slow-down in consumption, but also the growth of a new, large and low-cost producer, Brazil, which stood outside the ITA, and an expansion of exports from China and some smaller producers. The fall in demand as the communist regimes of Eastern Europe and the USSR collapsed, and some of their economies descended at least temporarily into chaos, contributed significantly to continuing low prices in the 1990s. This chapter first discusses the run-up to the sixth, and last, International Tin Agreement, a period which also saw an attempt by the Malaysians to manipulate the tin market. The chapter then looks at the ITA collapse and considers its effect on established producers and the reactions in those producing countries. It traces the rapid growth of the Brazilian tin industry in the 1980s, and finally looks at developments in the world tin market, including the activities of the Association of Tin Producing Countries, the influence of US stockpile releases and of sales from China.

THE RUN-UP TO THE SIXTH AGREEMENT[1]

The negotiations for the sixth International Tin Agreement (ITA6) were conducted in a rancorous atmosphere. ITA5 should have expired in June 1981. Negotiations for ITA6 started in May 1980. When this first conference, and several subsequent ones, failed to secure agreement, ITA5 was extended for a year. The United States, under the influence of the new Republican president Ronald Reagan, pulled out of the Agreement in June 1981.

There had been acute dissatisfaction among many producer and consumer members about the operation of the fifth ITA. On the producer side, dissatisfaction centred on the slowness of revision of the intervention price range for the buffer stock. For much of the late 1970s prices had stayed above the ceiling price. As the market weakened in the early 1980s with world recession, the intervention price range became of more practical importance to producers. According to information supplied to the ITC's Economic and Price Review Panel, which met every six months to consider costs and returns to investment, Bolivia's costs (excluding export taxation) were scarcely above the floor price even after a upward revision of the range in October 1981. Consumers, however, felt that tin prices were holding up well in comparison to those of other metals, and from 1977 to 1980 there had been four upward revisions of the price range, despite falling world consumption of tin (Kestenbaum, 1991, p. 34). Consumers also were irritated by producer country demands to include export duties and royalties as 'costs', when such taxation was under the control of producer governments. There was considerable consumer resistance both to export control and to the raising of the price range, especially on the part of the USA, usually backed by (West) Germany and the UK. One American official involved with the ITC, whom I interviewed in 1982, referred to the ITA as 'a cartel with consumer collusion' (Thoburn, 1982a, p. 230).

During 1981 the United States General Service Administration had been selling tin from its stockpile while the buffer stock manager had been engaged in support buying, despite an agreement dating from 1966 that the GSA would not disrupt the ITA's operations by stockpile sales. In July 1981 producer requests for an upward revision of the price range were rejected. From July 1981 an unknown group. later revealed to be the Malaysians, started buying tin on the London Metal Exchange. The cash (i.e. spot market) tin price rose to over £8000 a tonne in August, compared to an average of £6000 in January. The mystery buyer was operating heavily on the tin futures (three-month) market. Many other operators, thinking that the high prices could not be maintained, sold 'short' (i.e. sold for forward delivery tin which they did not possess), hoping to buy it more cheaply on the cash market when the time came for delivery. In November, the operations of the mystery buyer switched to the spot market. If the buyer could 'corner' a substantial part of the supplies of physical tin (i.e. tin on the spot market), transactors who had sold 'short' would have to pay high prices to buy tin with which to make the deliveries they were contracted to do on the forward market. By February 1982 there was a severe shortage of cash tin. A backwardation (an excess of the spot price over the three-months forward price) grew rapidly, peaking in mid-January at over £1000 a tonne, with a cash price of nearly £9000. As Kestenbaum (1991, p. 29) notes, the LME's rules clearly required it to avert a 'corner' at all times. Operators left 'short' were being squeezed hard, and to prevent bankruptcies the LME imposed a limit of £120 per tonne per day on the premium such traders could be asked to pay for cash tin to 'borrow' for later delivery. In other words there was a fixed penalty imposed for delivering late. As a result, there was a drastic fall in the cash price of tin, by nearly £2000 during a two-week period.

As much as 60,000 tonnes may have been bought by the buying group. At the time this was widely believed to be inspired by the Malaysians, but the Malaysian government did not officially admit its role until 1986; it conceded in response to pressure from Malaysian opposition politicians that it had made substantial losses in its attempt to corner the tin market.

ITA6 finally went into force in July 1982, having been delayed several times awaiting ratification from prospective members. In the event, not only the USA but also the then USSR and most of its satellites stayed out, as also did Bolivia. China and Brazil remained outside.

THE SIXTH AGREEMENT AND THE PRICE COLLAPSE[2]

Export controls were implemented by the International Tin Council in 1982 under ITA5, and remained in place at the time of the price collapse in 1985. By 1983 there was an approximate balance between supply and consumption; in fact Table 5.1 shows a slight deficit. However, large stocks overhung the market, with ITA6 inheriting 49,000 tonnes from the Fifth Agreement. Further tightening of export control (already at 64 per cent of base level – see Gilbert, 1987, p. 611) would have made difficulties for miners in producer countries. In any case further control might have been circumvented by smuggling, for example by the operators of the suction boats mining off the Thai coast who exported via Singapore. Thus the main task of ITA6, while maintaining supply–demand balance by export control, was to keep the stock overhang off the market. Yet the buffer stock, with its large inheritance of tin from the previous Agreement, was starved of cash resources, since the reduced membership of ITA6 had contributed only the cash equivalent in new subscriptions of 6000 tonnes of tin. The operation of the buffer stock was limited under ITA6 to a holding of just over 61,000 tonnes, including net forward purchases, and was limited in its borrowing by the need to use its tin stock as collateral (Anderson and Gilbert, 1988, pp. 6–7).

The buffer stock got into the position of having to hold off from the market a larger stock of tin than it was authorised to hold. In the event, not

FIGURE 6.1   Tin prices indices in sterling, dollars and Malaysian ringgit, 1960–90 (1960=100)

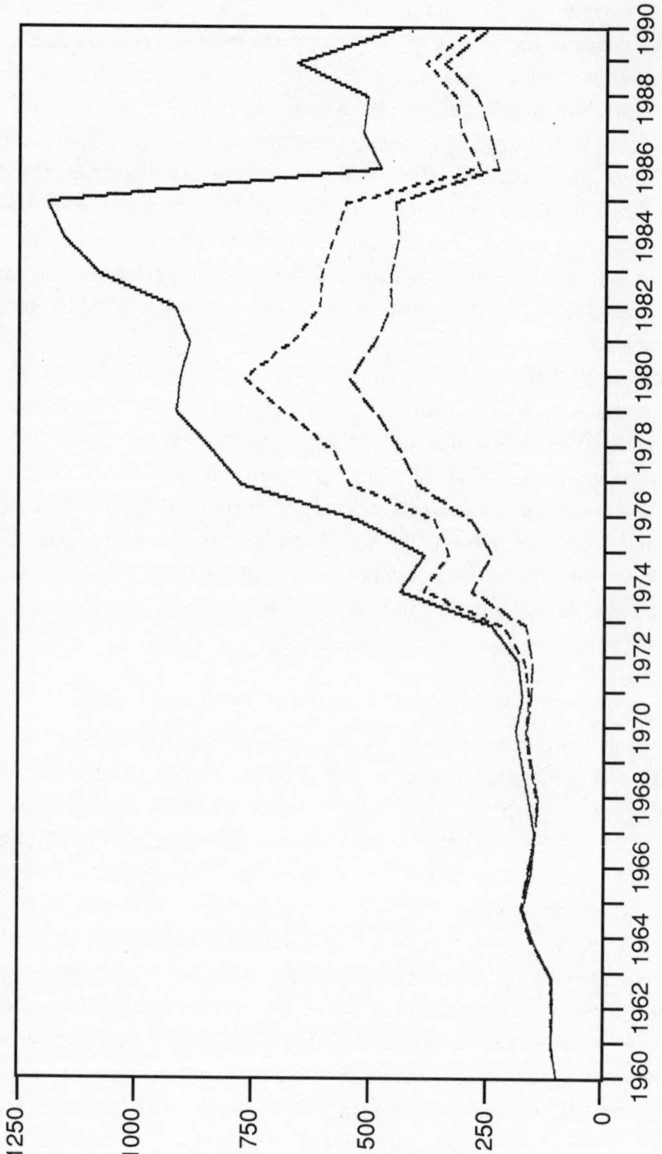

Sources and Notes

1. Prices are from various issues of *TI* and *MT*.
2. 1986–88 are European free market prices, the LME being closed.

LMEtinPI

NYtinPI

MtinPI

only did it fully utilise bank borrowing, but also engaged in several imaginative devices which allowed it to hold off a larger volume of tin than appeared officially as part of its holdings. These devices included 'special lends', where the buffer stock manager (BSM) would sell tin to a merchant while simultaneously buying it back forward at the same price plus interest; 'special borrows', where an agent would buy spot tin and simultaneously sell it forward, thus keeping it off the market; and 'unpriced forward sales', which could be deducted from the buffer stock's holding, but involved risk since they involved selling at an unknown price (Kestenbaum, 1991, pp. 38–9). The BSM's activities in the forward market for tin reflected the fact that he had insufficient resources simply to buy tin on the spot market and store it. His forward market operations also gave him more power against speculators.[3]

The operations of ITA6's buffer stock in the early 1980s were therefore a balancing act, involving active participation in the forward as well as the spot market. Forward purchases were matched with spot sales at maturity, so the BSM made profits if the market was rising and losses if it was falling. There could also be losses if cash tin was bought in Malaysia, and sold on the LME, since tin on the LME often traded at a discount to the price on the Malaysian market (Anderson and Gilbert, 1988, p. 9). The appreciation of the US dollar from mid-1982 to early 1985 had a strong effect on the buffer stock's profitability by raising the LME price, then denominated in £sterling, in relation to the Malaysian price, in which the buffer stock intervention range had been specified since 1972. As Figure 6.1 shows, the Malaysian Ringgit price of tin closely followed the US dollar price, which in the early and mid-1980s sharply diverged from the price in sterling. While the dollar continued to appreciate and the Sterling price of tin continued to rise, the buffer stock profitability was increased. When, following the Plaza agreement in 1985 to check the rise of the dollar a dollar depreciation started, there was a sharp decline in the ITA's financial position. On 24 October 1985 the BSM telephoned the LME to say he could not meet his obligations to pay for the tin he was contracted to take up that day, precipitating a default leaving a grand total of debts to traders and banks initially estimated at £900 million.

The LME suspended trading in tin, as did the Kuala Lumpur Tin Market (KLTM). Prior to the KLTM suspension, the spot price of tin stood at an average monthly price of Ringgit 29.91 (for October 1985). When it reopened early in 1986, the price had fallen to Ringgit 19.558 (average for February 1986), reaching a low for the year of Ringgit 13.99 in June.[4] When the Malaysian market reopened it amended its rules. Having previously been limited to trading in tin from Malaysian smelters, it allowed Thai and Indonesian tin to be traded also (Burke, 1990, p. 66). The LME did not resume trading in tin until mid-1989.

THE EFFECTS OF THE PRICE COLLAPSE ON ESTABLISHED PRODUCERS

Table 6.1 gives a detailed overview of the pattern of output from individual tin-producing countries over the decade to 1990. The decline of Malaysian, Thai and Bolivian output is striking, and Australian production also fell. This is true too of Nigeria and Zaire, though their outputs in 1980 already were small in relation to the past. Among the main traditional producers only Indonesia has maintained production. Equally striking is the rise in Brazilian output. Chinese output too is much larger than previously, though the change may to some extent reflect the collection of more accurate figures. There has also been some growth in output over the decade in Canada and Peru. This section first considers the effects of the price collapse on the South-East Asian producers. It then looks at Bolivia, and also at some minor producers, including the UK.

South-East Asia[5]

Table 6.2 shows the decline in Malaysian output and employment over the decade, using as a comparison 1972, when output was at its post-war peak.[6] The falls to 1980 reflect the gradual decline in Malaysian mining grades, compounded by various problems facing miners in prospecting for new deposits. The further falls to 1984 also reflect the ITA export controls introduced in 1982. There has been export control organised through the ATPC since 1987. Nevertheless, the fall in output also reflects the difficulty of many mines in covering costs at low prices. Reference back to Table 1.8 shows that at 1986 prices neither the Malaysian, nor the Indonesian, dredging or gravel pumping sectors *on average* could cover even their direct operating costs, and nor could the gravel pump sector of Thailand.

The effect of the price collapse has been to concentrate mining on higher-grade operations, while lower grade ones have been shut down or put on a care and maintenance basis. The greater falls in output in the Malaysian gravel pump sector compared to dredging are to be expected, the high price elasticity of supply of gravel pumping being based on a combination of low capital and high operating costs (Thoburn, 1977a). More interesting is that the relative fall in gravel pump employment is more than the fall in output, as reflected in the figures of output per worker. This is the result of a growth of large-scale operations and technical change in gravel pumping, as well as the closing of lower-grade operations.

Malaysian gravel pump miners since the mid-1980s have been adopting the technique of 'dry feeding', originally developed at the Gopeng mine in Perak in about 1970. This involves stockpiling the ore-bearing ground with earth-moving equipment to a area where it is then worked with gravel pumping, rather than having the gravel pumps work at the mine face. Dry-feeding allows much larger throughputs to be handled by a mine, so that

TABLE 6.1   World tin-in-concentrates production, 1980–90

(000s tonnes)	1980	1982	1984	1986	1988	1990
World	235.9	224.9	206.9	186.8	204.9	220.6
Developed market economy countries	19.3	21.4	16.7	18.5	15.7	17.4
America	0.3	0.2	0.3	2.5	3.7	4.1
Canada	0.2	0.1	0.2	2.4	3.6	4.1
USA	0.1	0.1	0.1	0.1	0.1	–
Europe	4.0	5.1	5.7	4.8	3.6	4.8
EEC	4.0	5.1	5.7	4.8	3.6	4.8
Portugal	0.3	0.4	0.3	0.2	0.1	0.5
Spain	0.4	0.5	0.4	0.3	0.1	0.0
UK	3.3	4.2	5.0	4.3	3.5	4.2
South Africa	2.9	3.0	2.3	2.2	1.4	1.1
Asia	0.5	0.5	0.5	0.5	–	–
Japan	0.5	0.5	0.5	0.5	–	–
Oceania	11.6	12.6	7.9	8.5	7.0	7.4
Australia	11.6	12.6	7.9	8.5	7.0	7.4
Developing countries and territories	182.2	169.0	151.5	122.8	138.7	141.4
America	36.2	37.2	43.5	44.0	59.5	61.9
Argentina	0.4	0.3	0.3	0.4	0.4	0.4
Bolivia	27.3	26.7	19.9	10.5	10.8	17.2
Brazil	6.9	8.2	20.0	27.7	44.1	39.1
Mexico	0.1	–	0.4	0.6	0.3	–
Peru	1.1	1.7	3.8	4.8	3.9	5.2
Africa (other Africa)	9.7	7.4	8.0	4.6	4.9	4.8
Burundi	–	–	–	–	–	0.1
Cameroon	–	–	–	–	0.0	0.0
Namibia	1.0	0.8	0.9	0.7	1.2	0.9
Niger	0.1	0.1	0.5	0.1	0.1	0.0
Nigeria	2.7	1.8	1.3	0.1	0.4	0.2
Rwanda	1.6	1.2	1.1	–	0.2	1.0
Uganda	–	–	–	–	0.0	0.0
United Republic of Tanzania	–	–	–	–	–	0.0
Zaire	3.2	2.2	2.9	2.5	1.9	1.6
Zambia	–	–	–	–	0.0	–
Zimbabwe	0.9	1.2	1.2	1.0	0.9	0.8
Asia (South and South-East Asia)	136.3	124.4	100.0	74.2	74.3	74.7
India	–	–	–	–	0.0	0.1
Indonesia	32.5	33.8	23.2	24.0	30.6	31.1
Lao People's Democratic Republic	0.6	0.6	0.6	0.4	0.2	0.1
Malaysia	61.4	52.3	41.3	29.1	28.9	28.5
Myanmer	1.1	1.6	1.9	1.4	0.5	0.5
Republic of Korea	–	–	–	–	–	–
Thailand	33.7	26.2	21.6	16.8	14.0	14.4
Least developed countries	6.6	5.7	7.0	4.4	2.0	3.4
Countries of Eastern Europe	18.0	18.0	19.7	19.0	18.6	15.8
Czechoslovakia	0.2	0.2	0.2	0.2	0.6	0.3
Germany (East)	1.8	1.8	2.5	2.8	3.0	0.5
USSR (former)	16.0	16.0	17.0	16.0	15.0	15.0
Socialist countries of Asia	16.4	16.5	19.0	26.5	31.9	46.0
China	16.0	16.0	17.5	25.0	30.0	44.0
Mongolia	–	–	1.0	1.0	1.2	1.2
Viet Nam	0.4	0.5	0.5	0.5	0.7	0.8

Source: UNCTAD (1992a)

TABLE 6.2   Malaysian employment and production, 1972–90

	1972	1980	1984	1986	1990
Number of mines					
Dredging	58	54	30	31	24
Gravel pumping	940	746	353	122	88
Employment					
Dredging	9,447	8,955	6,576	5,303	3,985
Gravel pumping	30,617	24,961	12,586	3,461	2,636
Total all methods	47,929	39,009	23,623	11,797	8,508
Output of tin-in-					
concentrates (tonnes)					
Dredging	23,989	18,222	12,728	12,090	10,267
Gravel pumping	42,800	34,484	21,577	11,146	12,216
Total all methods	76,830	61,404	41,307	29,134	28,468
Output per worker					
(tonnes)					
Dredging	2.53	2.03	1.93	2.28	2.58
Gravel pumping	1.40	1.38	1.71	3.22	4.63
All methods	1.6	1.57	1.75	2.47	3.35

Source: Figures supplied by Malaysian Ministry of Primary Industries, from Mines
        Department sources.

monthly throughputs of 200,000–300,000 cubic metres are not uncommon
among the few Malaysian mines still in business. The adoption of this new
technique in Malaysia was also helped by the fact that, as one miner in
Ipoh explained, after the 1985 price collapse, when many mines closed
down, there was a excess supply of contracting firms specialising in earth
moving (particularly dry-stripping) for the tin industry. The resulting competition between these contractors resulted in their rates falling far enough
to make dry-feeding more attractive. There has also been much greater use
of earth-moving equipment in Thai tin-mining than previously. In
Indonesia, in contrast, dry-feeding has been little used. Most Indonesian
gravel pump mines are quite shallow, and on the tin islands there is not
such a ready supply of contractors. Also, there has been a recent phasing
out of fuel subsidies, and earth-moving contracting in Indonesia is hampered by high import duties on the equipment.[7]

Increases in labour productivity in gravel pumping have not only been
due to dry-feeding, however. The manager of one large mine in South-East
Asia explained how his operation had more than doubled output per man
by a variety of minor improvements, and by some increases in capital intensity, since the price collapse. He stressed how much easier it was to

increase efficiency in times of low prices, when employees recognised the urgent need for change. One interesting aspect of this mine's improved performance was that it invested considerable time and energy in improving its relations with suppliers. Given that a gravel pumping operation consumes large quantities of spare parts, a regular and reliable supply of parts at competitive prices can cut costs and reduce down-time considerably. It would offer a long-term relationship to parts suppliers, with exclusive contracts in exchange for good prices, just-in-time delivery and priority as a customer.[8]

In Thailand a particular casualty of the contraction of the industry was the large sector of suction boats. These had operated on a quasi-legal basis since the late 1970s, but were still responsible for much of Thailand's smuggling of tin ore during the fifth International Tin Agreement. After the price collapse the Thai Department of Mineral Resources organised a high-ranking and effective task force to deal with the boats, and they have been replaced with a fleet of small suction dredges, also made locally. Some of these mini-dredges work as contractors to the Offshore Mining Organisation, which in the 1980s had been buying concentrates from the suction boat sector. The mini-dredges have proper moorings and their position can be checked more easily by the authorities than could the activities of the highly-mobile suction boats. As Table 6.3 shows, inland dredging has declined greatly. The apparently large expansion in offshore dredging is to some extent the result of the reclassification of the successors of the suction boats as offshore dredges. Thai gravel pump mining has contracted greatly over the late 1980s, as Table 6.3 also shows. Its output per worker has increased substantially since the mid-1980s, but is still far less than in Malaysian operations.

Another casualty of the 1980s was the offshore dredging operation owned by Billiton, which was allocated under the export control scheme a quota below its capacity. Billiton's large suction dredge was bought by a local company, which subcontracts its output to the formerly MMC-related Tongkah Harbour company. In the 1970s, as Chapter 5 showed, there remained in addition to Billiton only two foreign groups in Thailand, predominantly in dredging: one was related to the Malaysia Mining Corporation and the other to Fairmont State (43 per cent owned by the Cornish mining and property group St Piran). By the early 1980s MMC's involvement in tin-mining in Thailand had been reduced to minority ownership in two dredging companies: Aokam and Tongkah Harbour. MMC's stake in Tongkah Harbour was sold in 1985 to a local company (Roskill, 1990, p. 181), leaving MMC with only a 30 per cent share of Aokam (MMC, AR, 1991). The Siamese Tin Syndicate and Bangrin, owned by Fairmont State, had only one dredge in operation by the late 1980s. Fairmont State contracted out the management of dredging operations to a local company,

TABLE 6.3   Thai employment and production, 1976–90

	1972	1980	1984	1986	1990
Number of Mines					
Dredging inland	11	33 (inland + offshore)	28 (inland + offshore)	13	6
Dredging offshore	4	–	–	10	16
Suction boats	1986 (ITC figure)	2230	11 (number of mining concessions)	13	3
Gravel pumping	238	392	324	266	27
Open-cast	NA	NA	NA	110	37
Employment					
Dredging inland	2273 (inland + offshore)	3294 (inland + offshore)	3288 (inland + offshore)	1369 (inland + offshore)	730
Dredging offshore	–	–	–	–	1,517
Suction boats	11,308	33,540	3009	1401	111
Gravel pumping	15,518	15,626	9704	3944	1,151
Open-cast	7,869	NA	NA	NA	1,101
Total all methods	43,738	70,305	32,041	18,467	14,022
(Total excluding dulang washing)	(36,968)	(62,779)	(25,375)	(11,467)	(7979?)
Output of tin-in-concentrates (tonnes)					
Dredging inland	1872	4336 (includes offshore)	4995 (includes offshore)	1092	788
Dredging offshore	1459			2084	7175
Suction boats	4602	14,342	5218	3558	1
Gravel pumping	7822	9964	6534	4998	2498
Open-cast	1979	NA	NA	3225	2336
Total all methods	20,103	33,685	21,607	16,298	14,385
Output per worker (tonnes)					
Dredging inland	1.46 (inland + offshore)	1.32 (inland + offshore)	1.52 (inland + offshore)	2.32 (inland + offshore)	1.08
Dredging offshore	–	–	–	–	4.73
Suction boats	0.41	0.43	1.73	2.54	0.01
Gravel pumping	0.50	0.64	0.67	1.27	2.17
Open-cast	0.25	NA	NA	NA	2.12
All methods	0.54	0.54	0.85	1.42	1.80

Sources and Notes

1. Number of mines and output figures are DMR (1976) and DMR (1990), except for the 1976 number of suction boats, which is from ITC MSB Feb 1980; and 1980 and 1984 figures are from ITC (1986). Employment figures by method are from ITC MSB Feb 1980 for 1976 and from UNCTAD (1992b). The employment figure

and ceased to operate in Thailand as a company under its own name (Roskill, 1990, p. 174).

Although the dredging sector in Malaysia has maintained output since the collapse far better than gravel pumping, as has also the Thai offshore dredging sector, the tin industry in South-East Asia was in great difficulty because of the prices which prevailed for most of the second half of the 1980s and early 1990s. Many dredges in Thailand and Malaysia were only covering direct operating costs, and the newer ones were unable to put aside cash flow to cover depreciation. Routine maintenance was postponed in the hope of higher prices (which materialised only briefly in 1989). Malaysia Mining Corporation in its 1991 annual report recorded a loss on its mining operations, compared to a profit in the previous year, as a result of low tin prices. Out of MMC's total dredge fleet of 38, only 21 were in operation in 1989, and by the end of 1990 this had fallen to 16. The Group's policy was to maintain mining as its core business, but to improve tin productivity by keeping open only the viable units. MMC has diversified into alluvial gold projects in Malaysia and Australia, which to some extent call on its tin expertise, and has been prospecting for gold in New Zealand and Brazil, and also has prospected in other countries, including China. MMC has also diversified into engineering, and into manufacturing beyond its interest in tinplate production and smelting. Similarly, Selangor state's Kumpulan Peraangsaan Selangor (the organisation which controls the mining of the large, deep Kuala Langat deposit) noted in its 1990 annual report that 'the need for KPS to venture into new activities, especially in growth areas, has become more pressing to offset its high but non-remunerative investment in the tin sector'. Several Thai offshore dredging companies interviewed in 1992 reported similar problems, deferring maintenance and keeping going in order to cover cash cost in the hope of some recovery in the tin price once the overhang of stocks from the tin collapse has been cleared. The Thai

---

for 1986 is from a fact sheet from the DMR, and employment figures are not given in DMR (1976, 1990).
2. 1976 employment figures for open-cast may include some other minor methods. 1976 employment figures for gravel pumping include the minor method of hydraulicing.
3. 1976 is the first year for which statistics are given for suction boats.
4. DMR statistics for output are converted from tin concentrates at an assumed metal content of 72 per cent.
5. The output per worker figure for 'all methods' is slightly overstated since it excludes the (large) employment but includes the (small) output of dulang washers.
6. Note that there are significant inconsistencies between the various sources and even between different issues of the same source. The figures in this table should be treated with reservation.

Offshore Mining Organisation has operated its own offshore dredge since 1982. In the 1990s it was making losses on that operation. It maintained some income through royalties from contractors, though these by the 1990s were mostly selling direct to the Thaisarco smelter. Thaisarco itself was experiencing problems with throughput down from its 1980 peak of 35,000 tonnes to 12,000 in 1991, and was unprepared to make the new investment in modifications to the smelter which would be needed if it would import larger quantities of lower quality concentrates, as do the Malaysian smelters. Since the price collapse, Thaisarco has been able to market through the Kuala Lumpur Tin Market, but has actually been increasing its direct marketing in collaboration with its parent Billiton, which is still part of the Royal Dutch/Shell group. Smelters in the world tin economy generally been experiencing problems, as feed has reduced with the fall in tin output from many producers. Billiton has closed its smelter at Arnhem in the Netherlands (*MBM*, July 1992), and RTZ closed the Capper Pass smelter in the UK in 1991 (*TI*, February 1991); minor smelters in Europe, Korea and Singapore have also closed. Amalgamated Metal Corporations' smelter – Williams, Harvey – had ceased operations in the early 1980s. It had been operating under a liquidator since 1973, and suffered from a shortage of ore for smelting as domestic Bolivian smelting developed (*TI*, September 1981, June 1983)

Malaysians from the tin industry, as in the case of MMC, have shown interest in using their skills abroad. Delegations with large contingents of Malaysian miners, and with the help of the Ministry of Primary Industries, have been to a wide range of places with potential for mining, including China, Vietnam, and several Latin American countries. 'Reverse investment' has become a new buzzword in the Malaysian tin industry (*MT*, January 1993), as more miners look for overseas investment possibilities.

Although Malaysia Mining Corporation is the country's largest tin-mining group, and in Thailand the state Offshore Mining Organisation retains a small stake in the industry,[9] it makes sense nevertheless to expect the response of the tin industries of Malaysia and Thailand to the price collapse to be market-determined. Despite the Malaysian government's substantial stake, MMC is run as a profit-seeking, public listed company,[10] it operates its tin mines often with minority ownership of the individual operating companies, following the tradition of the London Tin Corporation, whose role it took over. P. T. Timah in contrast, despite its formal status as a limited company, is a state enterprise in the traditional mould. It pursues long-run profit maximisation, but this is subject to many social obligations, including the provision of social services (Radetzki, 1985, p. 76). Although Timah was highly profitable in the 1970s, with high tin prices (and rich deposits), it was also much less technically efficient than comparable Malaysian and Thai operations.

As Table 6.1 shows, Indonesia increased its tin output after 1985, and by 1990 it was producing as much as in the early 1980s, before its production started to be reduced by ITA export control. However, P. T. Timah recorded a loss in 1990 for the first time in decades (*Mining Journal*, 1991, p. 102), and in the late 1980s and early 1990s it became something of a flagship for World Bank restructuring of state enterprise. After a Bank mission to study the Indonesian industry in late 1985, prior to a loan, a master plan to restructure Timah was started in early 1990, commencing with a study lasting over six months by a major international consultancy company. One important recommendation was for Timah to concentrate on its core activities, and to divest itself of everything else.

In 1992 Timah was starting to implement major changes. It planned to reduce its overall labour force from 24,000 to 11,000; to cut the size of the dredge fleet, while producing the same output; to specialise on offshore tin-mining operations, leaving inland operations (mainly gravel pumping) to private sector contractors; to divest itself of the workshops used for the gravel pump operations; and to pass its social facilities (such as hospitals and schools) to the local governments in the islands in which it operates.

In 1992, nineteen dredges were operating offshore; four operated inland, all on Bangka island. At least half the dredge fleet was of pre-1970s vintage, some up to fifty years old. Timah aimed to concentrate production on its newer, high-capacity, offshore dredges.

For gravel pump operations, Timah has long had a history of using contractors, and the numbers have increased to over 300, some two thirds of whom are on Bangka island. Many of these are now independent operations, with their own equipment and labour force. They all sell to Timah at a fixed price which is the same within each area.[11] These contractors in the early 1990s were producing about a third of the output sold by Timah. In 1991 a new cooperatives organisation was set up, which in 1992 consisted on Bangka of forty-eight mine units. Each unit was a former Timah gravel pump mine, and under the cooperative it would employ only between a quarter and a half of the labour which it previously had used, though at the time the redundant workers were still being paid. The cooperatives worked with mining equipment being purchased from Timah on an instalment basis, and two-thirds of expected profits would be reinvested in the mine. The director of the cooperative spoke of diversification into other activities in order to employ redundant workers. Such schemes have included the planting of oil palms, but a scheme in the late 1980s to expand pepper production foundered when the world pepper price collapsed.

Timah's maintenance of its production level reflects the organisation's belief that, with its higher ore grades, it can compete effectively at low world tin prices only if it can reduce its operating costs (per unit of throughput) towards the level of those of Malaysia. The carefully-planned, gradual way in which Timah has been reducing its workforce reflects the very different employment circumstances in Indonesian tin-mining areas from those of Malaysia and Thailand. In Malaysia tin is mined near to the country's main industrial areas, and the rapid growth of the economy means alternative employment is readily available, especially for skilled workers. In Thailand, which also has a rapidly-growing economy, tin-mining is far away from the industrial areas, but it is in places where tourism has been booming in the 1980s and 1990s so, again, there is alternative employment. On Bangka island in contrast, which is Timah's largest producing unit, Timah is the largest employer. Its workforce is not large as a proportion of the island's working population, but the income generated is important as a source of cash receipts to the island's economy.[12]

What have been the reactions of governments in South-East Asia to the tin industry's problems, and what help have they given? In the 1970s there were many complaints in the tin industry about 'penal' taxation, especially in Malaysia (Thoburn, 1978b) and particularly about output-based taxes, which are the equivalent of additional costs to the mine. In fact, the sliding scale nature of the duties means that to a large extent the burden was removed automatically when prices collapsed. In Malaysia after 1985, miners rarely paid export duty since prices were too low for the duty to be triggered except briefly during the price rally in mid-1989; even then the duty would have been at very low rates. In late 1990 it was announced that the export duty would be abolished on tin and most other minerals in Malaysia, as far as the federal government was concerned, and taxation would simply be in the form of tax on corporate profits, but some export duty would be imposed by state governments at higher tin prices. During the 1980s the Malaysian government helped the tin-mining industry with a soft loan scheme, and tin miners also received a 25 per cent reduction in electricity charges. In Thailand in the years after the price collapse the royalty on tin production, already on a sliding scale, was reduced further at the lower end, though the reductions were not introduced as quickly as the industry would have liked (*MT*, third quarter, 1989). At the low prices prevailing at the beginning of 1992, the royalty took less than 1 per cent of the value of output, and a levy to pay for past buffer stock contributions took about another 4 per cent.[13]

In Indonesia, the royalty is imposed at only 2 per cent of the value of output at prices below $8000 a tonne [well above 1992–3 tin prices], rising to 10 per cent above $10,000. Otherwise taxation is based on corporation taxes. This applies both to Timah and to Koba Tin. Koba is now the only foreign investor remaining in tin, and is 75 per cent owned by Renison

Goldfields and 25 per cent by Timah. The underground mine operated by the Australian steel-maker and tinplate producer BHP at Kelapa Kampit was sold in 1985 to the German mining multinational Preussag, but Preussag sold its holding to the Indonesian company Gunung Kikara in 1990. The offshore dredging company P. T. Riau Tin, which was set up as a joint venture between the Indonesian government and Billiton in 1968, was liquidated at the end of 1985, affected by the price collapse and also the depletion of reserves (*TI*, January 1991).

Bolivia

Bolivia's reaction to the 1985 tin collapse has to be seen in relation to the fact that the country was already undergoing an economic crisis of its own. It faced difficulty, in common with many other Latin American countries, in servicing its substantial external debt in the face of rising real interest rates and recession in the world economy. Since tin was still an important source of foreign exchange (35 per cent in 1985), the solutions to the serious problems in the tin industry were an important part of any solution for the economy's difficulties more generally. Restructuring measures, in mining and elsewhere, were perceived as part of the conditions necessary to get International Monetary Fund and World Bank assistance.

In September 1985, the month before the ITA's demise, Comibol's mines were shut down by strikes against the wage freezes, removal of food subsidies and the administrative changes to be introduced in Comibol. These measures were part of the New Economic Policy introduced by the newly-elected government of President Victor Paz Estenssoro to restructure the economy – the same president, incidentally, who had led the government which took power after the 1952 revolution (Crabtree *et al.*, 1987). The measures included further devaluation of the Bolivian currency followed by the establishment of a freely-floating exchange rate.[14]

The poor performance of Bolivian tin in the early 1980s was against a background not only of Comibol's management shortcomings, but of the declining ore grades, lack of exploration and lack of mining investment which had characterised the industry since before the second world war.

From 1980 to 1984 Comibol's output fell by nearly a third, in part due to labour stoppages as miners protested against the various 'packages' of restructuring measures. During the early 1980s as metal prices and production fell, Comibol made large losses, although mineworkers' representatives have claimed that these were exaggerated by the government to justify mine closures (Crabtree *et al.*, 1987, p. 86).

Compared to Malaysia and Thailand, where workers after 1985 were able to find jobs in other sectors of the economy, and Indonesia where P. T. Timah seems to have at least some sense of social responsibility, the treatment of workers from Comibol has been harsh. This, of course, follows a long history of poor treatment of Bolivian mineworkers which continued into the period of nationalisation (Nash, 1979, 1992; Coote, 1992, ch. 2). Some 20,000 workers out of Comibol's workforce of 27,000 have been laid off with minimal compensation, and many workers in ancillary activities were also affected, as Comibol's highest cost mines were closed. Some have been able to form mining cooperatives to take over particular mines, but many have been left with little alternative but to migrate to the drug-growing areas in search of work, or to go panning for gold. In the mining areas there is little alternative employment. Comibol has been relieved of responsibility for providing subsidised food and the social infrastructure for its remaining workers, and in the cooperatives conditions are dangerous and workers poorly remunerated. Jordan and Warhurst (1992) argue that, in spite of these social costs, the measures have not been effective in changing Comibol's method of operation. The closures have not been associated with the introduction of new technology in the remaining mines, nor with any move away from the 'high-grading' of the richer deposits to maintain production in the absence of organisational improvements.

In 1990 new measures were introduced. These sought to encourage foreign investment and allowed Comibol for the first time ever to enter into joint ventures. These have included a twenty-year deal with the largest Brazilian tin-mining company, Paranapanema, to treat Comibol's large tin tailings dumps in the Potosi area, expected to yield nearly 3000 tonnes of tin a year (*TI*, July 1992). However, strong opposition by the mineworkers' union to the idea of joint ventures with the private sector continued, with a series of strikes (*TI*, December 1992).

Minor Producers[15]

By the 1990s Australia had 90 per cent of its tin production coming from the Renison Bell mine in Tasmania, owned by Renison Goldfields Consolidated, also the majority owner of Koba Tin in Indonesia. The Renison mine was temporarily closed in 1991 following disputes with the workforce over redundancies, as the mine had made losses. By 1992 the Renison mine was back in profit (Koba Tin was also profitable). Australia's main smelter, in Sydney, had closed in 1988, and has since been dismantled. A small smelter operated in Western Australia, but most of Australia's production was sent to Malaysia for toll-smelting. Domestic consumption of tin was equivalent to about a third of domestic mine production, and tinplate was the largest single use.

Nigerian output was by 1990 down to a mere 230 tonnes of contained tin metal. The Makeri smelter remained in operation, and was supplied mainly by one group, Consolidated Tin Mines, and there was minor production from the Nigerian Mining Corporation.

Zaire's production in 1990 still came mainly from the Sominki group, but Zairetain was nearing insolvency because of reserve depletion and the low tin prices. Although production in Zaire did not collapse to the same extent as in Nigeria, local production of concentrates was not enough to feed the country's smelter, at Manomo, and Zaire produced no tin metal.

In the UK there had been a minor revival of Cornish tin-mining in the 1970s and early 1980s as tin prices rose. This expansion was helped by the fact that the UK, since it was not self-sufficient in tin, was classed as a consuming country by the ITC and was therefore not subject to export control. In 1979 Rio Tinto Zinc bought the Wheal Jane tin mine, shut down at the time, from Consolidated Goldfields; it bought control in South Crofty in 1985, a mine in which Charter Consolidated had been involved (*TI*, July 1984). Until 1986 RTZ also had an 18.4 per cent stake in the third Cornish tin mine, Geevor (Crabtree *et al.*, 1987, p. 93). In addition, RTZ owned the Capper Pass smelter. Following the 1985 tin price collapse, the British government refused Geevor the financial aid the company sought, but offered loans to the two RTZ-controlled companies. It was generally believed that the British Conservative government's apparent generosity was influenced by the fact that it was challenged in that part of the country by the then Liberal/Social-Democratic Alliance, the Liberal Democrats of today, and feared for its electoral prospects if Cornish unemployment rose as a result of mine closures. By 1993 South Crofty was the only tin mine in operation. The Geevor mine had closed in 1986, though it reopened briefly as tin prices rose in 1988. In 1988 RTZ's share in Carnon, the company through which it owned South Crofty and Wheal Jane, was sold to a group of Carnon's managers and workers. After the government withdrew funding from Carnon in 1991, the Wheal Jane mine was closed, and untreated water escaping from mine has since been a serious environmental hazard. South Crofty remained Britain's only tin mine.

Canada's tin production had risen over the 1980s from negligible amounts to about 4000 tonnes a year due to the development of East Kemptville, a large open-cast lode mine. This was closed by its owners, RTZ, in 1992, leaving RTZ with the Neves Corvo mine in Portugal (which produced tin as a joint product with copper at low cost) as its only remaining source of tin. Neves Corvo, whose tin-copper deposit was discovered in 1986, was set up by RTZ in partnership with the Portuguese government, with RTZ holding 49 per cent and the government 51 per cent (*MT*, third quarter, 1989).

Myanmer (Burma) expanded its tin industry in the early 1980s with loans from the EC and the World Bank, and commissioned a dredge in 1980. A smelter was also established (*TI*, October 1986), but the later 1980s saw a large fall in production.

Peru, the third largest tin producer in Latin America, increased its tin output substantially during the 1980s, and by 1990 was producing more than the UK. The tin is produced by a private Peruvian company, Mansur, which purchased the largest mine from the American company W. R. Grace in 1977 (UNCTAD, 1990, pp. 50–5).

THE BRAZILIAN TIN INDUSTRY[16]

In the 1970s the industry already was developing in the Amazon region, with most of the same companies which are active today. Nevertheless, the potential for Brazil to become the world's largest tin producer did not then seem in prospect. The position changed with the discovery at the beginning of the 1980s of the Patinga deposit in Amazonas state, north of the Amazon's main city Manaus. Patinga was developed by Paranapanema, already the country's largest tin-mining group. Paranapanema started life as a civil construction company in 1961, and had acquired mining interests and exploration rights in the Amazon later in the 1960s. By the time of the 1985 tin price collapse, Brazilian tin production was already over 26,000 tonnes, of which Paranapanema accounted for some 70 per cent. Patinga's reserves have been estimated as high as 575,000 tonnes (UNCTAD, 1990, p. 38), enough for thirty-years' production at the mid-1980s level.

The surge in Brazilian output from 1988 to 1990, in part taking advantage of the temporary recovery of tin prices in early 1989, can mainly be traced to the discovery of the Bom Futuro deposit by garimpeiros at Ariquemes in Rondonia in October 1987. Brazilian tin production peaked at over 50,000 tonnes in 1989, compared to under 7000 tonnes in 1980.[17] Although garimpeiros had been banned from Rondonia Tin Province (which includes all of Rondonia, and parts of the states of Amazonas and Mato Grosso) by the Brazilian government in 1971, there had been some return of garimpeiros in the 1980s with gold discoveries. Because of recession in the economy and a growth in unemployment the government was unwilling to check garimpeiro activity, and they were accepted as a partial solution to a social problem.

In 1988 as part of a new Brazilian constitution, one article of the constitution gave priority to cooperatives of garimpeiros to work on areas where they were already mining. This was the first recognition of garimpeiro rights. Accordingly, on the discovery of Bom Futuro, the mayor of Ariquemes formed a cooperative of garimpeiros, *Coogari*, to operate the Bom Futuro deposit. Paranapanema acquired the company M. S. Minercao, which claimed some prior mining rights at Bom Futoro, and a protracted legal dispute started between the companies and the garimpeiros. Coogari tried to show it had been active in the area before M. S. Minercao put in its application for mining rights. The main Brazilian tin companies organised into a group, Ebesa [Empresa Brasileira de Estanho SA], in which Paranapanema had the leading position, and which included Cesbra, Best, Brumadinho, Cia Industrial Fluminese and SNA (*MB*, 3 September 1992).

The Bom Futuro deposit is a boomerang-shaped hill of tin about 1600 metres in length and 500 metres wide. The 'garimpeiros' working on the hill generally use trucks and excavators, rather than the simple manual methods that traditionally have characterised garimpeiro activity. Initially the tin is found within a few metres of the surface, though these easily-accessible deposits have now been worked. Established companies complain of the environmental damage caused by garimpeiro mining, where tailings are dumped in the rivers. Environmental damage from small-scale tin-mining, though, is less than is the case with gold, where mercury is used and subsequently dumped. In 1988 18,000 workers were said to be mining at Bom Futuro (*MB*, Tin Supplement, 1988, p. 23).

In the course of the dispute between the established tin companies and the garimpeiros about Bom Futuro, the government made an appeal for the two sides to work together, and a formula was agreed whereby the companies would buy 80 per cent of garimpeiro output for 55–60 per cent of the London Metal Exchange tin price, and be free to bid for the remaining 20 per cent. Garimpeiro output rose from some 17,000 tonnes in 1988 to nearly 24,000 in 1989, falling to about 12,000 in 1990 and 10,000 in 1991.[18]

In other words, in 1988–9 the Brazilian garimpeiros alone were producing substantially more than the whole of Thailand! Tin industry estimates suggest that some 19,000 tonnes of garimpeiro output have been exported illegally from Bom Futuro 1989–91 (*MBM*, July 1992) via Bolivia. Bom Futuro is near the Bolivian border, and the area is important for the drug traffic, with tin being exchangeable for drugs.

The expansion of garimpeiro output in the late 1980s was an important factor in keeping down the tin price after the 1989 rally. In 1990 and 1991 Paranapanema recorded losses. In part these were due to problems in its construction business, but since 80 per cent of its turnover came from tin, the tin price was the dominant factor (*MB*, 16 April 1992)

Control over mining rights at Bom Futuro has been in continuing dispute. In September 1992 there was a court order for work to be halted, and the garimpeiro cooperative Coogari obtained a ruling overturning the authorisation given to Ebesa by the DNPM the previous year (*TI*, October 1992). After September 1992 work ostensibly remained halted by the Rondonia state court pending its decision as to whether Ebesa or the garimpeiros should have the mining rights. In fact, work continued at the site! At the beginning of 1993 it was reported that monthly output was about 1200 tons of tin concentrates, with about 2000 garimpeiros operating side by side with Ebesa, and selling concentrates to Ebesa (*MB*, 1 February 1993, cited *MT*

January (sic) 1993). This output is equivalent to an annual production of tin metal of over 7000 tonnes.

The Brazilian tin industry is now even more concentrated than in the 1970s. In 1991 Paranapanema alone produced 58.6 per cent of Brazil's output of tin-in-concentrates (Paranapanema, 1992). Other major companies were Cesbra, Brumadhino and Best, all in existence in the 1970s. Of the multinationals operating in Brazilian tin in the 1970s, Patino has pulled out. British Petroleum bought 50 per cent of Cesbra from the Canadian company Brascan in 1981, but BP sold its stake back to Brascan in 1989 as it moved out of mining. The main mining companies all have smelting interests. Paranapanema also manufactures mining equipment.

THE TIN MARKET SINCE 1985

As we've seen, the sixth International Tin Agreement was brought down by a continued overhang of stocks beyond the resources of the buffer stock manager to hold off the market, exacerbated by the price effects of exchange rate changes. The end of 1985 saw approximately 100,000 tonnes of tin stocks overhanging the market (Table 6.4), including 26,000 tonnes of minehead stocks of tin-in-concentrates (ATPC, 1986).

The formation of the Association of Tin Producing Countries was announced by Malaysia, Indonesia and Thailand in June 1982 as the fifth agreement was coming to an end, and was inaugurated formally in September with a membership of Malaysia, Indonesia, Thailand, Bolivia, Nigeria, Zaire and Australia. Of the other five countries 'eligible' to join (i.e. the world's other net exporters of tin), two major producers (Brazil and China) remained outside; the others were Burma, Rwanda and Niger (ATPC, 1983), all very minor producers (see Table 6.1). While the ITA remained in being, the ATPC served as a forum in which producer positions could be formulated. Once the ITA had collapsed, the ATPC assumed the role of supply control.

Some ATPC documentation is critical of 'a certain large consuming country (i.e. the USA), which ultimately did not even join the Agreement' for exerting pressure to dilute ITA6's economic provisions, such that ITC export control was 'too little too late' in the face of a build up of tin stocks (ATPC, 1985). Nevertheless, the general tone of ATPC material is decidedly unconfrontational. Export control was agreed between ATPC members, to start from March 1987, but the ATPC stressed (1986, p. 28) that the object of this 'Supply Rationalisation Scheme' was to remove excess stocks from the market, not to cartelise:

> Thus, whilst the supply rationalisation is not meant to jack up prices to unsustainable levels, its implementation would help to underpin the tin price around levels sustainable by free market forces.

'Excess' stock is the amount above a 'normal' stock level of about 20,000 tonnes (a little over 10 per cent of annual world demand). The original stated expectation of the ATPC was to achieve a reduction in stocks to this level in two-three years (i.e. by about the end of 1989) (ATPC, 1986, p. 12), after which export control could be removed.

Compared to ITA6, the producer membership of the ATPC restriction scheme is large. Bolivia joined, and there has been some degree of agreement between the ATPC, Brazil and China. Brazil, China and the ATPC together in 1987 controlled about 85 per cent of world tin supply. Compared to the 39.2 per cent production cutback under ITA6 from 1983 to 1985, the first ATPC supply restriction scheme involved cutbacks of only 7.7 per cent compared to estimated production (ATPC, 1986, p. 24). However, the price fall by itself has been a powerful force making for production cutbacks, and some ATPC members have not been able to produce as much as their permissible export tonnages.

Reference to Table 6.4 shows a substantial cut in stocks through to 1988. In early 1989 there were significant rises in price, and at one point the price rose above the floor price of ITA6 before the collapse, reaching RM29.15 in the Kuala Lumpur Tin Market on 17 April. The Supply Rationalisation Scheme foundered in late 1989 and 1990 on increases in production from non-members. Chinese sales increased from 7000 tonnes in 1985 to around 19,000 tonnes, despite repeated assurances by China that it would contain production. Moreover, as several South-East Asian ATPC participants mentioned in interviews, exports claimed by the Chinese and cited in Chinese official export statistics indicated a level of sales only about half that recorded in importing countries trade statistics.[19]

The increase in Brazilian production, a major factor in the ITA's demise, continued into the late 1980s, and was greatly exacerbated by increases in garimpeiro output from the Bom Futuro deposit (see earlier). Tin smuggling by garimpeiros, mainly through Bolivia, is reflected in the 13,000 tonnes of tin of 'unspecified origin' in Table 6.4. ATPC annual reports record efforts by Brazil to control garimpeiro production, and such control is in any case greatly in the interests of the established Brazilian tin companies led by Paranapanema. Decex, the department of external trade in the ministry of the economy, has required since May 1990 that tin comes from a registered source, which excludes garimpeiro production other than that sold to Ebesa members (*TI*, November 1990). Ebesa members, in turn, are subject to limitations on exports agreed between Brazil and the ATPC. Some smuggling continues, and reports in 1992 indicated that the Bolivian authorities, in talks with the Brazilians, would not acknowledge that the problem even exists (*TI*, March 1992).

Table 6.4 also shows sales from the US strategic stockpile continued after

TABLE 6.4   World supply and demand for tin, 1985–91

(000 Tonnes)	1985	1986	1987	1988	1989	1990	1991
Supply: production of tin-in-concentrates							
Indonesia	21.8	24.9	26.2	29.6	31.3	30.2	28.7
Malaysia	36.9	29.1	30.4	28.9	32.0	28.5	20.7
Bolivia	16.1	11.0	9.2	10.5	15.7	17.2	16.8
Thailand	16.6	16.8	14.8	14.0	14.7	14.4	10.6
Australia	6.9	8.7	7.7	7.3	8.1	7.2	5.7
Zaire	3.0	1.9	1.9	1.9	1.8	1.7	1.3e
Nigeria	0.9	0.8	0.2	0.6	0.4	0.2	0.3e
(Subtotal: total ATPC)	102.2	93.2	90.4	92.8	104.0	99.4	84.1
Brazil	26.5	25.8	28.5	44.0	45.0	39.6	29.5e
Others	19.2	18.5	18.6	17.5	18.8	21.1	23.3e
Unspecified origin	11.0	2.5	NA	2.0	13.0	6.5	8.3e
(Sub-total: fresh tin production)	158.9	140.0	132.9	156.3	180.8	166.6	145.2
US stockpile disposals	3.0	5.5	4.1	2.4	2.8	2.2	6.2
Chinese sales	7.2	5.4	16.6	19.0	18.7	15.9	15.7
Total supply (before net change in stocks)	**169.1**	**150.9**	**153.6**	**177.7**	**201.6**	**184.7**	**167.1**
Demand: consumption							
EC	38.3	45.3	44.6	50.7	50.0	51.6	50.3
USA	37.1	32.5	40.9	37.0	37.1	36.9	36.3
Japan	31.6	31.5	31.8	32.2	33.8	34.2	35.3
Other OECD	9.9	7.4	6.9	7.1	6.9	6.9	6.5
ATPC	7.4	7.8	7.4	8.9	8.9	9.4	12.0
Brazil	NA	5.8	5.5	7.8	8.3	6.1	6.5
Others	22.0	21.4	21.0	21.7	22.5	21.9	23.3
Eastern Europe	11.4	12.7	11.8	11.5	11.4	9.9	6.7
Imports into former USSR	14.1	14.0	14.5	11.0	11.2	2.4	0.0
Total demand	**171.9**	**178.4**	**184.5**	**187.9**	**190.0**	**180.6**	**176.9**
Total Ending Stocks	104.5	76.3	45.8	35.0	45.0	49.0	39.1

Sources and Notes
1. All figures from ATPC (1986, 1987, 1989, 1990 and 1991).
2. 1985–8 figures for former USSR include imports into former German Democratic Republic.
3. 1987 'Others' figures appear to include tin of 'unspecified origin'.
4. There are some minor discrepancies in the figures. For example, production figures for Bolivia, Australia and Nigeria for 1986 differ between the 1986 and 1987 ATPC annual reports.

the tin collapse. In most years since 1985 the stockpile has sold more tin than the combined production of Nigeria and Zaire, though of course stockpile sales were small compared to some years in the 1960s and 1970s (refer to Table 5.1). Run by the American General Services Administration until the late 1980s, in the 1990s the stockpile was operated by the US Defence Logistics Agency. As part of the disposal of surplus stocks of forty-four commodities, the US Congress in 1992 authorised disposal of 141,278 tonnes of tin (*TI*, December 1992). More immediately, producers were worried by the authorisation for the DLA to dispose of 60,000 tonnes of tin over a five-year period (*Bangkok Post*, 28 February 1992).

The dispute between the International Tin Council and its creditors was finally settled by the ITC membership on 30 March 1990, with most creditors getting back about 35 per cent of their losses (Kestenbaum, 1991, p. 175). One result of the settlement was that some 7000 tonnes of tin stocks became available to the market in 1990 (ATPC, 1990).

Supply restriction by the ATPC has continued into 1993, though the 1993 supply restriction scheme involves a slight increase in the permissible export tonnages over 1992. Brazil agreed to lower exports from 28,000 to 24,000 tonnes, and China agreed to maintain exports at 15,000 tonnes. China has continued to express intentions of joining, though it has not done so. Prices in mid-1993 were down to below RM14/kg on the Kuala Lumpur Tin Market, with tin stocks building up, recession in OECD countries, and the continued collapse of tin imports into the former USSR which started in 1990.

OVERVIEW: THE 1980S AND 1990S

The 1980s saw the start of a rapid decline in real tin prices. Nominal tin prices fell sharply in terms of US dollars (and in Malaysian ringgit, in which the ITA buffer stock intervention range was set), but initially sterling tin prices rose because of exchange rate changes. Most other minerals had been experiencing price falls as world demand fell in the recession following the second OPEC oil price shock in 1979. The ITA attempted to arrest the decline by sharp export control, but was left with the task of absorbing larger stocks of tin than it was authorised to hold. It was able to continue to hold these stocks off the market for several years as the US dollar appreciated, causing the LME price in sterling to rise. The sudden fall of the dollar after the 1985 Plaza agreement on international exchange rates caused the ITA great financial difficulty, and led to its defaulting on its obligations at the London Metal Exchange. Thereafter, the tin price went almost into free fall. Over the decade the tin price fell by more than that of any other major primary commodity. This was at a time when primary commodities generally were perceived as being in a state of crisis, with prices down to the real levels of the 1930s depression. Although the Association of Tin Producing Countries introduced export control after the ITA's collapse, the ATPC's activities were undermined by the continued rapid growth of exports of tin from Brazil, which overtook Malaysia to become the world's largest producer. China, like Brazil, stood outside the ATPC, and greatly increased tin sales by China were an additional factor leading to lower prices. After a brief rally at the end of the 1980s and the start of the 1990s, tin prices continued downwards, helped on their way by continuing Chinese sales and the collapse of ex-Soviet import demand.

In retrospect, although the proximate fall of the ITA can be traced to adverse exchange rate movements, the underlying cause clearly was the development of new output sources outside ITA control. This was combined with a long-term loss of market share to other materials, probably worsened by the rising real tin prices of the 1960s and 1970s. Yet tin's commodity agreement had lasted for a longer period than that of the four other post-war ICAs (cocoa, coffee, natural rubber and sugar) which had economic provisions (i.e. the ability to intervene in their markets).[20] The ICAs for all five commodities had been conceived at times when the market environment was much less difficult than the 1980s, and all aimed primarily at price stabilisation. Tin was distinctive in the dramatic nature of its collapse, but the coffee agreement broke down in 1989 (over disputes about the allocation of quotas), and the sugar agreement lost control over the market in 1982, after which its economic provisions lapsed. The sugar and cocoa agreements have been judged to have been almost totally unsuccessful at stabilising prices (Gilbert, 1987, p. 613).

Of the five ICAs, only tin combined buffer stock operations with effective provision for export control. Cocoa and natural rubber used buffer stocks alone, and coffee and sugar had export control agreements. Tin, with the Economic and Price Review Panel of the ITC, saw acrimonious disputes between consumer and producer members about the price range. Some consumers, especially the Americans, clearly regarded the ITC as being virtually a cartel. The Americans left the fifth (and penultimate) ITA, the only one of the tin agreements they had joined, having failed to persuade members to rely solely on the buffer stock mechanism. In some other commodities, the Americans successfully had insisted on using only buffer stocks for stabilisation.[21] The ITAs in the post-war period had to live under the threat of US stockpile releases of tin. These releases, or the possibility of them, had been the main way that the ceiling price was maintained, though there certainly were periods too when the ceiling was breached. However, the 1985 collapse cannot be blamed on the relatively small US stockpile releases in the 1980s.

The effect of the price collapse on Malaysia and Thailand was severe, particularly on the gravel pumping sectors, whose profitability is highly price sensitive. In both countries gravel pumping employment by the end of the 1980s was but a small fraction of that at the beginning of the decade, as all except the mines with the richest deposits closed. Dredges normally have much lower direct operating costs than gravel pumps, and the decline in dredging output and employment, though severe, has been less than in gravel pumping. Neverthess, many dredges too were put on a care and maintenance basis, and the Malaysia Mining Corporation, the world's largest single operator of tin dredges, made losses on its tin-mining operations. In Indonesia it was felt that the industry was better able to survive low tin prices than Malaysia or Thailand, but the state tin company P. T. Timah found it necessary to undertake a major programme of restructuring. This included divesting itself of many ancillary activities and making greater use of mining contractors and cooperatives. In Bolivia the state tin corporation, Comibol, had been in severe difficulty long before the tin collapse, and the Bolivian economy was undergoing severe structural adjustment, including a large real devaluation just before the ITA's collapse. This real devaluation has been sustained and, together with measures to privatise many of Comibol's operations, may help make Bolivian tin competitive, but at heavy social cost to mineworkers already with wretchedly low standards of living.

Most of the large multinationals who entered the industry in the 1960s and 1970s had left by the 1990s, and most of Western smelting capacity closed due to lack of concentrates to process.

The favourable effects of tin-mining on economic development in South-East Asia set out in Chapter 5 have been difficult to sustain. The low real prices have meant the loss of mineral rents both to local entrepreneurs and shareholders, and to government in the form of lost tax and export duty revenue. There has been great loss of employment. In Malaysia and Thailand there have been growing alternative employment opportunities in the mining areas, so that there has been a relatively smooth transition to other activities. A similar transition has been achieved by some, though not all, of the backward linked activities, particularly engineering, which have found other customers. In Indonesia, although the economy has been growing rapidly, the geographical concentration of mining on the three tin islands makes the redeployment of entrenched workers more difficult. In Bolivia, tin's development effects have never been strong, and the restructuring of Comibol has left many workers with little alternative but to move to the cocaine-growing areas.

NOTES

1. This section draws mainly on Burke (1990), Jomo (1990) and Thoburn (1982a). It also uses material from the fascinating first-hand account of the tin crisis by Ralph Kestenbaum (1991), a negotiator who represented one of the ITC's major creditors.

2. This section again draws on Burke (1990) and Kestenbaum (1991). Anderson and Gilbert (1988) give a useful economic analysis of the collapse of the Agreement, particularly its relationship to the tin futures market, and Gilbert (1987) is a good source of information on ITA6's operation. The House of Commons trade and industry committee reports on the tin crisis also contain much detailed information (HoC, 1985–6, 1986–7). For a formal analysis of the effects of exchange rate changes on commodity prices, see Gilbert (1989).

3. Burke (1990, p. 61), however, notes that in June 1985 when speculators were caught short and a substantial backwardation developed, the LME suspended trading for a day and levied only a small penalty on traders unable to deliver tin to meet their forward sales commitments. This was in spite of the fact that the BSM had offered to deliver tin at a price such as to maintain orderly business.

4. Figures supplied by Kuala Lumpur Tin Market, March 1992.

5. This subsection draws mainly on interviews in South-East Asia in February and March 1992. Roskill (1990), which discusses the tin industry in the late 1980s, has been a useful cross-check on the interview material, especially for Thailand.

6. Indeed, this output was surpassed only by production in 1940 and 1941, and only in two other years (1929 and 1937) was more than 70,000 tonnes produced (ITSG, 1949).

7. In the 1970s diesel and other petroleum products could be obtained in Indonesia at roughly half the world price. These implicit subsidies were phased out during the 1980s as part of the structural adjustment of the economy (Thoburn, 1981a, pp. 150, 155; Tambunan, 1990).

8. This company was in fact using the subcontracting techniques, though not explicitly saying so, made familiar by Japanese industry and now adopted by many Western industrial companies (Thoburn and Takashima, 1993).

9. In September 1992 *Tin International* reported that the Thai government planned to privatise OMO, which the government criticised as having been poorly run.

10. As of 1991, the largest shareholder in MMC, with 43.6 per cent of MMC's shares, was Permodalan Nasional Bhd, the Malaysian state investment trust. There was also 6.93 per cent held by PNB's unit trust and marketing arm ASN – Amanah Saham Nasional Bhd (See MMC's annual report, 1991). Charter Consolidated, MMC's partner in the acquisition of control of the London Tin Corporation, sold its stake in 1987.

11. All contractors on Bangka island are paid the same price, and all contractors on Belitung are paid the same price, but that price is higher in Belitung, reflecting Belitung's lower ore grades.

12. The *total* population of Bangka is 650,000, and Timah in 1992

employed 12,727. On Belitung island in 1990 Timah's workforce was 7770, with a further 42,000 dependents, out of a total population of 160,000. Employment in the Singkep mining unit in 1990 was 2230 (*TI*, December 1990). Heidhues (1992, pp. 212–13) estimates that, including dependents, a quarter of the population of Bangka in 1985, and a third of Belitung's, depended for their livelihood on the tin industry. She argues, however, that pepper, because it is grown so widely on Bangka, has more of an influence on income on the island than tin.

13. Figures supplied by Thai Mining Industry Council, February 1992.

14. It is worth repeating here that, with regard to the inter-war period of tin control, Hillman (1988a) has argued that Bolivia should be able to restore international competitiveness in tin by devaluation, and that a depiction of Bolivia as a high-cost producer would be misleading. This possibility arose because Bolivian exports were dominated by tin. In other tin producers, where tin was a much smaller proportion of foreign exchange earnings, the exchange rate could hardly be determined by the export demand for tin (see Chapter 3). Since tin in the mid-1980s still dominated Bolivia's export earnings – or at least its legal ones – the principle could still hold. This issue is taken up in Chapter 7.

15. Unless otherwise indicated, material in this section draws on ATPC ARs for 1989 and 1990, and various issues of *TI*. See also HoC (1985–6, 1986–7) for much detail of the effect of the tin collapse on British tin-mining.

16. This section is based mainly on interviews held in Brazil in August and September 1992.

17. Paranapanema (1992) gives the 1989 output figure for Brazil as 54,637 tonnes of tin-in-concentrates. This differs from the ATPC figure in Table 6.4 because that table does not attribute any of the 'unspecified' tin to Brazil. UNCTAD (1992a) gives a 1989 figure of 50,200 tonnes.

18. Figures are estimates from SNIEE, the Sindicato Nacional da Industria da Extracao do Estanho, supplied at interview in Rio de Janeiro in September 1992.

19. China's exports of 'tin and tin alloys' for 1989 are listed in SSB (1991) as 9874 tonnes, compared to the ATPC figure in Table 6.4 of 18,700, though there is some possibility that the SSB figures do not include exports of tin concentrates for smelting overseas.

20. There were also informal arrangements for jute and sisal. The formal ICA for jute (from 1983), and for tropical timber (from 1984), contained exclusively 'development' provisions (such as measures to improve productivity) and did not aim at price stabilisation (Maizels, 1992, pp. 135, 137). Maizels (especially chs 7 and 8) gives a very useful account of international commodity policy. The present discussion draws on this source, and on Gilbert (1987).

21. This applies to the 1979 natural rubber agreement, and the 1980 cocoa agreement (Maizels, 1992, p. 137).

The Goose that Laid the Golden Egg
by Redzwan Sumun, December 1985

Requiescat in pace, Tin Agreement,
Though this is just yet but a premise,
For many, with little lament,
Have augured thy untimely demise.

Thy life of but twenty-eight years,
Has been filled with chequeredness,
In the assuage of thy members' fears,
Of tin price's dreaded unevenness.

Many a time thou hast been put to light,
As an example of the full consumation,
Of nations' hope that it is a right,
In the north to south cooperation.

Thou hast modified with skill consumate,
Man's natural greed to gamble,
His endowment, his precious estate,
And thus prevent many a tumble.

But over time thou got taken for granted,
That in thyself alone thou can cope –
In the end finding thyself unwanted –
Sans aid, sans nurturing of scope.

Being nought but man's creation,
Thou hast need for man's management-
But this need is cast to oblivion,
Notwithstanding his early sacrament.

Hence thy success is thy own undoing,
Thy golden egg raped to lead,
By those that stand most benefiting,
In the earnings of the daily bread.

Thou now languor in the throes of death,
And in the horizon barely is sympathy,
While thou draws thy last pained breath,
Friends and foes look down in apathy.

While success may breed success,
Breeds it also many an enemy –
And in the stupor of many a caress,
Was hidden the stab of fatality.

I salute thee thy noble intent,
Thy tenacity in facing adversity;
To thinking men thou have so lent
The unlearnt concept of universality.

In thy demise I feel the stress,
And tears of regret my soul takes –
Man learns not from his success,
And less still from his mistakes.

CHAPTER 7

Conclusions

The world tin industry changed rapidly between the mid-nineteenth century and the first world war, as modern industrial demand for tin replaced traditional uses. The old-established dominance of the British tin-mining industry in Cornwall was successfully challenged by the growth of new sources of supply. Colonial Malaya became the world's largest producer, and Bolivia became important too. Britain ceased to be the largest consumer of tin, and the United States tinplate industry developed rapidly. While Britain had strong colonial interests in tin, not only in Malaya and (indirectly) in Siam, but later in Nigeria and Burma too, the Americans had no significant domestic tin production. British interests also had a large stake in smelting, which was started in Malaya even in the nineteenth century, and much Bolivian production was smelted in Britain. British colonial policy actively supported their nationals' tin interests, as did that of the Netherlands with regard to the then Dutch East Indies. This happened first in the formation of the international tin cartel in the years between the two world wars, and, in Britain's case, with regard to the exclusion of the Americans from the smelting industry. These conflicts would leave the Americans ill-disposed towards the efforts of independent producing countries, after the second world war, to control prices through the International Tin Agreements.

While the economic structure of the tin-mining industry changed after the second world war, with nationalisations in Bolivia and Indonesia, and with ownership restructuring in Malaysia and Thailand in the 1970s, the geographical pattern of production established by the first world war remained largely intact until the 1980s. Then, the growth of a large new producer, Brazil, encouraged by the high real prices of the 1970s, undermined the activities of the International Tin Agreement, already weakened by a secular decline in tin consumption. The ITA's dramatic collapse in 1985 started a fall in tin prices which, at the time of writing (1993), has continued downwards with only a temporary respite in the late 1980s. This has led to fears for the future of tin-mining in much of South-East Asia and Bolivia.

This chapter now picks out three themes to conclude the book: the changing economic structure of the industry, especially with regard to the role of multinational companies in LDC producers; the development effects of tin-mining; and the future of tin in the light of the collapse of the International Tin Agreement.

CHANGES IN INDUSTRIAL STRUCTURE

The international structure of the industry which had been built up in the late 1920s and early 1930s, centring on the London Tin Corporation and Patino/Consolidated Tin Smelters, survived in most important respects into the 1960s and even the early 1970s. Although Patino and the other tin 'barons' were dispossessed as a result of the Bolivian nationalisation in 1952, Patino's powerful control over the smelting of Bolivian ore was not broken until Bolivia developed its own smelting operations in the 1970s.

These LTC and CTS groups were 'multinational' in the sense that they operated in several countries, but they were quite specialised on tin. In the 1960s and 1970s the large, diversified international companies which were predominant in the mining of most other metals started to enter tin. This was part of a more general trend towards new entry into metals, and of cross-penetration between activities, which also saw the proliferation of state-owned enterprises in the minerals industry. Thus Charter Consolidated, run by the South African group Anglo–American, bought heavily into the London Tin Corporation companies, and also acquired tin mines in the UK and Portugal. Rio Tinto Zinc, the world's second largest mining corporation after Anglo–American in terms of the value of its mining output (Roskill, 1992, p. 13), entered Malaysian tin with a dredging operation. RTZ later bought into the Cornish tin-mining industry, and set up large-scale mining in tin in Canada. It also acquired a British smelter. Consolidated Goldfields, a company with South African connections, developed tin in Australia and the UK, as well as owning tin mines in South Africa. Billiton, nationalised in Indonesia in 1958, diversified into lead and zinc. It also moved back into tin in Indonesia with an offshore dredging investment in 1968, into offshore dredging in Thailand, and had an unsuccessful foray into tin prospecting in Brazil. Billiton developed smelting in Thailand in the 1960s with the Thaisarco smelter, and ran the Arnhem smelter in the Netherlands. The oil majors, who diversified into the mining of other minerals, also entered tin. Royal Dutch Shell acquired Billiton in

1970, and during the 1980s British Petroleum owned a stake, with the large Canadian mining company Brascan, in the Brazilian tin company Cesbra.

Thus the economic structure of the tin-mining and smelting industry by the early 1970s was already rather different from that in colonial times. In the 1970s, the two largest producers where foreign-controlled production remained important, Malaysia and Thailand, started to behave more nationalistically. Ownership restructuring was forced on the tin companies. Although their local ownership had increased anyway since the second world war as a result of local share purchasing, these enforced changes brought local control. In Malaysia, though, this was exercised with the South African controlled multinational, Charter Consolidated, in partnership with the new Malaysia Mining Corporation.

Multinational involvement in the tin industry, already weakened by the early 1980s, was greatly reduced first by ITC export control quotas, and then by the 1985 price collapse. Billiton ceased mining in Indonesia and Thailand. Charter sold its share in Malaysia Mining Corporation in 1987.[1] RTZ, though it was still expanding into tin in the early 1980s, eventually divested itself of almost all its tin operations. BP divested itself of its tin interest, though Royal Dutch Shell has retained Billiton. A few successful operations remained, such as Renison Goldfields' Koba Tin in Indonesia. Goldfields itself, after a takeover battle with Anglo–American in the late 1980s, became part of the Hanson group and maintained its interests in tin in Australia and South Africa.

Much of the multinational involvement in tin has been replaced by state-owned enterprise (SOE). Most of the Indonesian and Bolivian tin production is carried out by SOEs, as of course is also the case in China. Malaysia Mining Corporation is majority-owned by state agencies, Thailand has the Offshore Mining Organisation, and Nigeria the Nigerian Mining Corporation. In the tin industry, however, unlike many minerals, there has been a viable local sector in most producing countries. This has put a limit on the growth in concentration in many producers, although one of the most famous international tin companies, Patino, was a local Bolivian company which grew worldwide. The small/medium scale technique, gravel pumping, provided local firms in Malaysia and Thailand with an opportunity to compete against larger scale foreign companies operating dredges, though the 'locals' running these mines were mainly immigrant Chinese. Only in Indonesia was the industry largely in the hands of the state even in colonial times. The Brazilian tin industry is predominantly in the hands of indigenous Brazilian companies, though there has been acrimonious competition with very small-scale miners.

Tin smelting has suffered greatly from overcapacity, especially with the falls in world output after the 1985 collapse. Smelting in Malaysia has imported concentrates worldwide, to compensate for the decline in Malaysian mine output. Thaisarco operated at about a third of capacity in the late 1980s. Consolidated Tin Smelter's successor, Amalgamated Metal Corporation, bought out by the German minerals company Preussag, maintains a stake in one of the two Malaysian smelters (Datuk Keramat), but its Williams, Harvey smelter in the UK closed for lack of feedstuff before the tin price collapse. Other smelters, including Billiton's Arnhem smelter, have closed too. The vertical integration between mining and smelting, developed by Patino and LTC, remains in some respects. Bolivian smelting is run by the state, and Comibol and private Bolivian mines must sell to the state smelters. Most Brazilian mining companies own smelters. However, Preussag does not own Malaysian mines, and recently sold its one mine in Indonesia. The other Malaysian smelter, the Straits Trading Company, though it had some long-term investments in tin-mining, essentially acted independently of mining interests, but was required to join with Malaysia Mining Corporation to form the Malaysian Smelting Corporation to operate the smelter.

MMC also has stakes in Malaysia's single tinplate manufacturer, and in local mining equipment supply. In that sense there has been more vertical integration in the industry in Malaysia than in the past. However, in general vertical integration in tin remains slight beyond the smelting stage. Tinplating remains an adjunct of steel manufacture, and is carried out by a small number of producers such as the British Steel Corporation, though local tinplating has been set up in other producers such as Thailand and Brazil. Solder manufacture is little integrated with tin, except that some solder is produced by the smelters in Malaysia and Thailand.

In other metals, there is a tendency now for the economic nationalism which mineral-producing countries in the Third World showed in the late 1960s and 1970s to be reversed. Such countries are now more likely to welcome multinationals, recognising the MNCs' potential contribution to economic efficiency and easier access to external finance. On their side, the MNCs are perhaps less rapacious than in colonial times, more willing to join joint ventures (including ones with state-owned companies), and more willing to act generally as good corporate citizens.[2] In the tin industry, most new entry by MNCs took place just before, or during, the start of a period of nationalism. Most exited in the 1980s, first as ITA exports cut permissible outputs (Billiton in Thailand for example), and after the collapse (e.g. Rio Tinto Zinc's exit from Cornish tin). There has been virtually no new entry. State-owned enterprises in tin have started to undergo painful periods of adjustment, as in the case of P. T. Timah in Indonesia and Comibol in Indonesia. Timah has divested itself of many activities, and the fashionable remedy of privatisation is being tried on Comibol. Thailand's Offshore

Mining Organisation too may be privatised. Malaysia Mining Corporation, though majority-owned by Malaysian government agencies, has behaved more like a commercial company than an SOE.

THE DEVELOPMENT GAINS FROM TIN EXPORTING

The development effects of tin exporting comprise those which have occurred through market forces, and those which have been brought about by policies in the exporting countries. In tin, the gains which have been secured by policy consist mainly of tax receipts, income from the restructuring of ownership (including nationalisation), and some attempts to promote tin-using industries. In the case of tax and of ownership restructuring, short-run gains may in principle have to be offset against longer-term losses; investment, including foreign investment, may be discouraged. International competitiveness may be damaged too, for example, if state-owned enterprises prove unable to compete at world prices. In the case of policy-induced 'linkages' such as local tinplate manufacture, a case needs to be made that the activity is economically beneficial. Only the development effects in Third World producing countries have been considered in this book. These producers generate the bulk of world tin output.

Malaysia's experience illustrates most clearly this book's contention that tin exporting has been unusual among mineral exports to the extent that it has generated favourable effects on development.[3] An important aspect of these beneficial effects has been the development of mining techniques, particularly gravel pumping, which could be used by local miners in the face of competition for mining deposits by foreign investors. The smallness of tin deposits compared to those of many other minerals has helped too to reserve a place for local miners. Gravel pumping, which was still the world's most widely used method in the 1980s, is of relatively high labour intensity (compared to Malaysian manufacturing for example). It has generated labour skills among its workforce; it has a high ratio of net to gross foreign exchange earnings (i.e. high 'retained value'), and it buys much of its spare parts and some capital equipment from the local economy (high 'backward linkage'). Cost benefit studies have shown it to be highly socially profitable (Thoburn, 1977a). Even the more capital intensive placer mining technique, dredging, at least in the 1970s (i.e. before the expansion of Malaysia's labour-intensive manufactured exports) was no less labour-intensive than domestic manufacturing. It required a high proportion of skilled workers, and bought mining supplies and capital equipment locally. Tin-mining was instrumental in developing the Malaysian transport system, and was a key factor in spawning a local engineering industry. Broadly similar effects have occured in Thailand, though the mining areas were remote from main centres of population. In the nineteenth century, powerful immigrant

Chinese helped Thailand to maintain its independence from Britain, and were involved in the setting up of the first locally-owned modern tin smelter in Malaysia. In Indonesia, where the industry was mainly developed by the colonial state, similar mining techniques were used, and immigrant Chinese comprised the bulk of the workforce, but wages and working condtions during the colonial period were poor compared to those of Malaysia. There was much less local capitalist development in mining in Indonesia, though there was some contracting to Chinese mine operators.

In Nigeria however, like South-East Asia a producer of alluvial tin, the only major study of the industry (Freund, 1981) has argued that tin-mining generated little development. His case seems to rest on the argument that tin-mining greatly disrupted the local African economy. It caused labour to be withdrawn from agriculture in a coercive way (e.g. by taxation which forced Africans into wage employment). There was actually forced labour during the second world war. Also, there was very little participation by Africans other than as workers; small-scale techniques were not enough to provide a niche for them in the face of other obstacles. Freund concedes there was some local development in activities servicing the mines. Since the economy is far less developed industrially than Malaysia or Thailand, a lower degree of responsiveness to linkage opportunities is to be expected, though in Malaysia especially, some of the early industrial development itself can be traced to tin. Freund also indicates that the tin industry, like other Nigerian exports, was damaged by the 'Dutch disease' effects of the Nigerian oil boom in the 1970s. With declining ore grades and lack of investment (despite MMC's involvement through its takeover of the London Tin Corporation), Nigerian tin output was falling before the collapse of tin prices in the 1980s; the low prices have accelerated the industry's decline. In South-East Asia, disruption of the local economy along Nigerian lines did not occur because most tin-mining was developed by overseas Chinese who eventually stayed as immigrants (though indigenous miners had been in evidence at an earlier stage). In Malaysia and Thailand, Western foreign investors did not become well-established until the early twentieth century. That Africans were prevented from being mine owners, and forced to become mine workers, shows that neither the influence of technology, nor economic effects more generally, work in a political and social vacuum.

Malaysia progressed from being a tin-export economy in the late nineteenth century to a more diversified primary commodity exporter, first with rubber, then palm oil, and in the post-war period with timber. It has since moved into export manufacturing, especially electronics. Thailand and Indonesia never were mainly dependent for their export earnings on tin, but had a wider range of primary commodity exports. Indonesia had become a significant petroleum producer even before the second world war.

Thailand more recently has joined Malaysia as an exporter of manufactures.

None of the South-East Asian economies, then, has become a classic 'mineral export economy', heavily dependent on minerals for both export revenue and national income, locked into mineral exporting by 'Dutch disease' effects via its real exchange rate, and stunted in its growth in other sectors. Bolivia, in contrast, has remained a mineral economy, based on tin, although tin has been supplemented by cocaine exports. The poor conditions in Bolivian mines have continued to the present. Nationalisation of the main private tin-mining companies in 1952, inheriting a legacy of decades of underinvestment, seems to have purchased additional employment at the cost of reduced international competiveness. Depreciating the exchange rate to reverse this trend, as argued below, would be (and has been) at the cost of the further erosion of living standards. In any case, Bolivian mine deposits are more difficult to work (and to smelt) than the alluvial deposits of South-East Asia and Brazil. Attempts to divert resources from Comibol to other sectors, though sensible in principle, have starved Comibol of essential funds for reinvestment.

Nationalisation of the remaining foreign interests following Indonesian independence after the second world war was also followed by cumulative falls in output, but then by a substantial investment programme. Indonesia, like the other South-East Asian producers, benefitted from a world market price which was nearer to the higher cost level of the lode mines of Bolivia.

The more nationalistic policies of Thailand and Malaysia in the 1970s certainly secured for them a larger share of the mineral rents generated at a time of high tin prices. In those two countries, and to a lesser extent in Indonesia, the share of export duties/royalties in relation to mining costs rose substantially in the course of the 1970s (Robertson, 1982, p. 34), in addition to which there were taxes on profits. Also, as I have argued elsewhere on the basis of investment appraisals (Thoburn, 1981a, ch. 6), the level of taxation was not sufficient in Malaysia and Thailand to discourage new investment, given the very high tin price.[4]

Tin-producing countries by the 1970s had been more able to secure a share of the mineral rents than was seen with most other minerals (Hughes and Singh, 1978). In part this has been due to the absence of control over the tin market by major mining multinationals. Even though major multinationals had control over as high a share of tin processing (i.e. smelting) facilities as in other major metals (Girvan, 1987, p. 717), smelters tended not to try to exercise market power themselves. Patino's behaviour in using his control over smelting to seek compensation from the Bolivians for nationalising his mines is an exception. Hennart (1986b) has argued, using a transactions cost framework, that Patino integrated forward into smelting because of his potentially poor bargaining position *vis-à-vis* the very small

number of smelters capable of smelting his ore; the investment requirements for smelting ore produced from lodes being much higher for alluvial
tin. Integration between tin-mining and smelting has continued in Bolivia
in the era of nationalisation, though with the smelting being set up within
the country. However, even Patino did not integrate into tin-using industries, these activities being technically dissimilar from smelting and mining.

Forward integration by tin-producing countries into tin-using industries
has developed since the 1970s, and earlier in the case of Brazil. Domestic
production of tinplate takes place in Malaysia, Thailand and Indonesia, and
has accompanied the development of local steel industries. Although in the
1970s it was fashionable to see 'resource-based industrialisation' as an obvious
way in which primary exporting LDCs could secure for themselves a greater
share of the gains from their primary production, there are grounds to be
sceptical about its benefits where it has not developed through market
forces (Roemer, 1979; Wall, 1988). Tin is a tiny fraction of even the cost of
tinplate, let alone of a final product such as a can of drink,[5] or of an electronics
product using solder. While smelting has the clear advantage of a 25 per cent
weight loss, as concentrates are turned into metal, the economic viability of
tin-using industries depends more on the cost of other inputs. Malaysia protected the growth of domestic tinplating by a 20 per cent tariff (Meyanathan,
1988, p. 209). To the extent that such an industry uses imported sheet steel,
as it did in Thailand (Praipol, 1988, p. 304), the nominal tariff may understate
the effective rate of protection if the steel products or other raw materials
have a lesser duty. To the extent that tinplating uses inefficiently-produced
local steel,[6] the economic efficiency of tinplate manufacture is damaged.
Further research would be necessary to determine the desirability of such
developments, but it is appropriate to raise reservations here. However, in the
case of South-East Asian countries which are significant producers of canned
fruit, an export product is available into which local tinplate production, if
efficent, could be incorporated. Similar considerations apply to solder used
for export electonics manufacturing in the South-East Asian producers.

Tin-producing countries have also exercised more influence on world
market prices than is the case in most other commodities, achieving what
Baldwin (1983) has dubbed 'political pricing'. Before the second world war
major Western tin-mining interests allied themselves with colonial governments to restrict exports and raise the price by, effectively a cartel. The
British government also tried hard to protect British interests in tin, particularly to prevent the Americans from challenging British smelting interests
before and during the second world war. Post-war American hostility to the
International Tin Agreements to some extent can be traced to the United
States' experience in facing the tin cartel and the British government protecting its colonial tin interests.

Since the price collapse in 1985, and with even lower prices in the early 1990s, the mineral rents in South-East Asia have largely vanished, and taxation has been greatly reduced. The legacy of tin-mining in the form of trained labour and linked industries remains. Much of the engineering industry serving tin (though not all of it) has been able to turn to other activities.

There has not been enough research time in the writing of this book to study the development effects of the Brazilian tin expansion. Unlike other LDC tin producers, Brazil developed tin smelting ahead of mining, using imported concentrates. Its early development of steel facilitated the development of tinplate, albeit with protection from import competition. Brazilian tin-mining is located in the Amazon, far from the main centres of population. Major mining multinationals have not played a large part in the industry, which is now mainly in the hands of a few large, indigenous firms. The main firms are integrated forwards into smelting, and there is some backward integration into mining equipment

THE PRICE COLLAPSE AND THE FUTURE OF TIN

In the 1970s tin might have been seen as a mineral nearing the end of its 'life cycle'.[7] As an increasingly scarce metal, it was experiencing historically high prices which were cutting consumption down to available supply. The largest producer, Malaysia, had falling output in the mid and late 1970s despite the high prices. Yet, by the early 1980s, considerable ITC activity was necessary in order to support the tin market. The 1985 price collapse was the result of a secular slowdown in tin consumption, which had been stimulated to some extent by the high real tin prices of the 1970s; together with the rapid expansion of tin-mining in a large new producer, Brazil. Brazil's expansion also was encouraged by the high prices. Chinese sales of tin on the world market were a contributory factor too. The International Tin Agreement through its use of export control and buffer stock purchases was able to maintain the price of tin for longer into the 1980s than other major metals maintained their prices (Crowson, 1987). In the event, the default of the ITA was due as much to adverse exchange rate movements as to the real underlying factors of increased production and decreased demand. In the late 1980s tin prices recovered temporarily, at a time when other metal prices were also strengthening. The late 1980s saw a substantial rise in world tin consumption, so the recovery was not simply due to the activities of the Association of Tin Producing Countries' supply rationalisation scheme. After the late 1980s rally, prices again started to fall.

Price forecasts have been made in a paper by Peter Frame (1992) for the ITRI and in the World Bank's (1990) *Commodity Trade and Price Trends*. Frame provides details of alternative low/medium/high consumption estimates, and uses US Geological Survey estimates (i.e. Sutphin *et al.*, 1990) for reserves and costs. He calculates that existing reserves have a life of between thirty and thirty-four years (depending on the consumption estimates). His low consumption estimate assumes a slight secular increase in world consumption, and the higher estimates make more optimistic assumptions about the demand for tinplate and about imports into the former Eastern bloc. Frame uses cost data compiled from the World Bank (1990), from ATPC and ITC sources, and from some consultants reports. Both the World Bank and Frame expect a long-term price (in constant 1989/90 dollars) of about $7000 per tonne. This compares to average New York prices of $6490 in 1986, $6,660 in 1987, $7040 in 1988, $8470 in 1989, $6072 in 1990, and $5456 in 1991 (converted from $/lb, from Crowson, 1992b, p. 264). The World Bank expects the price of $7000/tonne to prevail by the late 1990s, and Frame comments that it should last some twenty years, while there remain relatively 'low cost' reserves to exploit (i.e including not only Brazil, but also South-East Asia, but not Bolivia).[8] The World Bank stresses the role of potential increases in production from marginal mines in South-East Asia in keeping prices to this expected level. Producing secondary tin from recyling would require a price of around $9000/tonne to make it viable (Frame, 1992, p. 22). These estimates essentially are based on assuming a long-run price which will cover the costs of the South-East Asian producers as well as Brazil's. Frame cites direct mining costs per tonne of metal as $6300 for Malaysian and Thai gravel pumps, and $6700 for their dredges, and $5500 and $5800 for Indonesia for gravel pumping and dredging, respectively.[9] Smelting costs are estimated at $500 per tonne metal. Brazilian costs for mining placer deposits range from $3600 to $4400, while lode mining costs are $6200 for China and $9000 for Bolivia. Clearly Brazil occupies a pivotal position in determining the world tin price, and substantial increases in production by Brazil could lower prices significantly. These could occur, for example, if there were again uncontrolled garimpeiro production on a new deposit, as in the 1980s.

Tin's low prices in the 1990s, down again in real terms to the levels of the 1930s depression,[10] are in part a reflection of recession in OECD countries, as well as an increase in production in 1988–9. Some special factors also have been involved. Although Brazilian output fell by 1992 to less than half its 1989 peak (WBMS, 1993), China continued to sell tin, as it had in the 1980s. A new and significant factor was the effect of the breaking apart and economic collapse of the Soviet Union. Eastern bloc consumption of tin fell by over 10,000 tonnes between 1989 and 1991, with only a 2500 tonnes fall in production; Soviet imports fell to zero (UNCTAD, *ITS*, July 1992).

Leaving aside Chinese sales and the fall in Eastern bloc imports, consumption trends in the early 1990s in some respects seemed more

favourable than some for some time, representing a continuation of trends
of the late 1980s. Western world consumption, though it fell slightly in
1992, was still substantially higher than in the early and mid-1980s (WBMS,
1993). There was scope for growth in demand for both tinplate and solder
in the newly industrialising countries of Asia. A revived Russia in the
future might also generate a large demand for packaging material, as much
of its food output is lost as a result of the absence of packaging. The 'tin-
intensity' of tinplate was thought by commentators to be reaching a minimum limit,[11] or at least its rate of decline was greatly slowing down. The
World Bank (1990) argued that tinplate's share of the packaging market
seemed to have stabilised after years of decline, although there remain enor-
mous differences in the relative shares of aluminium and tinplate drinks
cans in different European countries (Frame, 1992, p. 14), so there is scope
for changes in particular markets. More generally, and at the margin, a low
real price for tin will help maintain market share.[12] However, tin-free steel
remains a competitor to tinplate, especially in Japan, and steel-makers obviously have a long-term commitment only to maintaining their sales of steel,
not to whether the particular coating should be chromium (as in TFS) or
tin. Tin's prospects in solder (now its largest single use), despite the rapid
growth of the electronics industry, have been lessened somewhat by technical change, especially surface-mount technology in the assembly of printed
circuit boards (World Bank, 1990, p. 130). On the other hand, the tin content of lead/tin solder for plumbing has increased with public concern
about the effects of lead on the safety of water supplies, and there also has
been pressure in the US to reduce the lead content of solders in electronics
(EIU, 1991, p. 39). The demand for tin chemicals has risen.

Events in 1993, however, seem for the moment at least to have made the
earlier price forecasts irrelevant, and the possibility is raised of a smaller
world tin industry than hitherto known. By mid-1993 the KLTM spot price
had fallen below RM13, lower than the lowest price (RM13.99) in 1986 after
the ITA's collapse.[13] The chief executives of both Malaysia Mining
Corporation and Indonesia's P. T. Timah called for the abolition of the
ATPC, arguing it had 'failed' and was 'useless' (TI, July 1993). Most striking
of all, Malaysia Mining Corporation, the inheritor of the London Tin
Corporation's tin empire, announced it was pulling out of tin-mining, after
three years of losses on its tin-mining operations (TI, May 1993). At the
mid-1993 LME price of $5000,[14] only Brazil could have covered its direct
costs, although of course individual South-East Asian mines may still have
been viable. Indeed, even at 1991–2 prices Brazil's tin exports themselves
were discouraged by a lack of profitability (MBM, July 1992), and some
Brazilian smelting capacity was closed (EIU, 1991, p. 36).

Of course, a country's mining 'costs' are affected by its exchange rate.

For most producers, where tin is but a small percentage of total export earnings, the real exchange rate is determined by flows of traded goods other than tin (and by capital flows). Where a fall in the price of tin is accompanied by falls in the price of other (more important) export commodities, the exchange rate may depreciate anyway. This raises the domestic currency receipts for tin mines too (equivalent to lowering their costs measured in foreign currency). This has happened in the case of Indonesia, whose main export is petroleum, where the price also fell both in the mid-1980s and more recently. Hillman (1988a), in work on Bolivia during the inter-war period, argued that Bolivia was only notionally a high cost producer; the dominance of tin in its export earnings meant that the exchange rate could be adjusted if the tin price fell. For example, in the case he was considering, a price fall could have come about if there had been economic warfare among the participants in the inter-war tin cartel. In fact, Bolivia continued to have serious problems after the tin collapse, even though it had had, just before the tin collapse, the largest exchange rate depreciation compared to any major tin producer in the year following.[15] These difficulties of course spring in part from Comibol's poor management, and from its history of investment starvation, which long antedated the 1952 nationalisation. They also reflect the fact that devaluation, as argued earlier,[16] is not a painless way of adjusting the balance of trade. In order to achieve a depreciation of the real exchange rate, real wages often have to be cut. Such cuts are particularly unpleasant in a situation such as Bolivia's, where mining workers are seen by almost every commentator as having wretchedly low living standards already. In any case, Comibol's cumulative lack of investment is likely to mean that it also lacks supply capacity to expand output in response to depreciation.

Clearly, if these exceptionally low prices continue, much of the industry's existing capacity, like that of Malaysia Mining Corporation, might be taken out of production. There is some scope for individual mines to mine only their richer orebodies and even to increase output in order to lower unit costs (Crowson, 1987). Many South-East Asian mines have already made great efforts to cut costs and to reorganise to improve efficiency, for example by the use of earth-moving equipment on gravel pump mines, and by better supply chain management with regard to the purchase of spare parts. Further efforts are difficult in the face of negative cashflow. In the case of Malaysia, production of tin was falling even in the mid and late 1970s, when prices were very high in historical terms. This decline, which characterised gravel pumping more than dredging, was less the result of high taxation of mineral rents than of difficulties in securing mining leases. By the 1980s and 1990s, various famous Malaysian tin companies had left tin-mining and were engaged in other activities such as property development.

Malaysia had in the early 1990s been addressing difficulties about mining land tenure through work on a new national minerals policy. If tin prices recover to the $7000/tonne level predicted by the World Bank, Malaysian dredges which have been put on a care and maintenance basis could be restarted. In Malaysia, the specialist engineering firms who in the past have fabricated and erected tin dredges mostly remain in business, doing other types of structural steelwork, and could turn to rehabilitating dredges. In the meantime, however, some dredges may be sold for scrap or to mine other minerals (such as gold), and sea-going dredges which have been decommissioned, as in Thailand, deteriorate rapidly through corrosion. In the gravel pump sector, generally regarded as having a high price elasticity of supply,[17] revival may be more difficult than is supposed. A miner interviewed in the mining state of Perak in Malaysia in 1992 explained that much of the vital supply infrastructure of the region had been lost. Contractors who could erect mining structures such as palongs, with little need for instruction from the miner, had gone out of business, as had many of the foundries who produced mine spare parts. Skilled men had left the industry, and the area, for employment elsewhere. Tin in the 1990s was widely regarded in Malaysia and Thailand as a 'sunset' industry, and it seems unlikely that the downward trend in production will be reversed in the face of declining ore grades and other difficulties as it was, for instance, in the US copper mining industry (Crowson, 1992a). In Malaysia, tin deposits historically have been near centres of population, and the major cities which originally developed as mining towns have been competing with the tin industry for urban land. In Thailand, although tin-mining is far away from the main centres of population, it is in a major tourist area where the environmental effects of tin-mining are unacceptable and where onshore mines may have to compete with hotels for prime sites. Since both the Thai and Malaysian economies have been booming in other sectors, resources are relatively easily reallocated out of the mines. Thus, the structure of the world tin-mining industry may see further change as its focus shifts from two of the three main producers in South-East Asia. China's future role is more difficult to assess; there is little information about costs, or about how export decisions are made. Certainly one may expect further changes in the geographical pattern of production, but there may be some consolidation of long-term demand if real prices remain low. Malaysia and Thailand now seem ready to consign much of their tin-mining to economic history, and to shift resources into other sectors of their growing economies. Parts of the industry may survive there, especially if prices stabilise at higher levels than the early 1990s, and more of the Indonesian industry is likely to stay in production than in Malaysia and Thailand. Obviously, the removal of production capacity from Malaysia and Thailand improves the prospects for those producers remaining in the industry. In the case of the large South-East Asian producers, one may conclude that tin has contributed more to their development over the years than would have most other colonial-style exports of comparable foreign exchange earnings

The ITC itself is unlikely to be resussitated. Its statistics-gathering functions have been taken over by UNCTAD, but even modest proposals to set up an International Tin Study Group to take over from UNCTAD (UN, 1990a) have met with resistance.

NOTES

1. Charter also, in 1993, bought out the 36 per cent stake owned by Anglo–American's company Minorco (*The Guardian*, 24 June 1993), thus shedding its most obvious South African connection.
2. These arguments have been put forward, for example, by Crowson (1991) and Radetzki (1992). This is part of a more general welcoming back of MNCs into the Third World (see Harvey, 1991).
3. For a discussion of the effects of other minerals on development, see Thoburn (1977b, chs 10 and 11).
4. Thoburn (1981a, ch. 6) also has some discussion of the relative merits of taxation versus government equity participation in tin investment projects as the best means of securing a share of the mineral rent for the host economy. I argue that the choice depends both on the discount rate used and the projected tin price; hence no general conclusion emerges.
5. In the 1970s, tin cost 0.5 per cent of the cost of a beverage can, and 3.6 per cent of the cost of the tinplate (Robertson, 1982, p. 83). Since then the coating of tin on tinplate has thinned further, and, of course, tin is much less expensive.
6. McKern (in McKern and Praipol, 1988, pp. 372–4) argues that import-replacing steel industries in South-East Asian countries tend to be uncompetitive at world prices. Studies of ASEAN countries and Australia in Findlay and Garnaut (1986) give data on effective rates of protection in various manufacturing industries: with regard to tin-using industries, only the study of Malaysia (by Lee Kiong Hock, p. 131) gives an ERP figure for any tin-using industry. Tin cans and metal boxes were protected by a 15 per cent nominal tariff in 1973 and 1978, and though the effective rate of protection fell between those two years (from 76 per cent), it was still 33 per cent in 1978.
7. See Humphreys (1982) for a discussion of the mineral 'life cycle' concept. He is sceptical of it. He stresses the role of price in regulating the relation between production of a mineral and the resources of that mineral, and sees the relationship as essentially economic rather than physical.
8. Remember though that the US Geological Survey admits its figures greatly understate Brazilian reserves (see Chapter 1).
9. No explanation is given by Frame of why gravel pumping's 'variable costs' should be lower than those of dredges, which seems surprising. Possibly it is simply a typing error, with dredging and gravel pumping transposed. Alternatively, it may be assuming that

only the most efficient gravel pump mines with the richest deposits remain in being. Frame later (p. 22) suggests adding an average 10 per cent to mining and smelting costs to cover 'overheads'. The World Bank (1990, pp. 132–3), in contrast, cites a Malaysian government survey of tin costs in 1987 which gives costs as $6/kg ($6,000/tonne) for gravel pumps and $5.2/kg ($5,200/tonne) for dredges.

10. See Figure 1.1. The average real price of tin in 1992 (not shown in Figure 1.1) rose slightly compared to 1991 (1991 was 25.6, 1992 was 27.7, with 1980=100). Even the very low price of $5,000/tonne in mid-1993 (index approximately 22) was slightly above that of 1931 in real terms. Reference to Figure 1.1 also shows that there were times in the late nineteenth century when real tin prices were even lower than in the 1990s and the inter-war years. However, in the late nineteenth century South-East Asian countries were working exceptionally rich deposits (like Brazil's in the 1980s) which could yield profits at real prices low in today's terms.

11. Frame (1992) and World Bank (1990) provide useful commentaries of the future of tin consumption, on which the present account draws. See also EIU (1991).

12. I am unaware of any recent studies of the price elasticity of demand for tin. Robertson (1982, p. 86) cites a study estimating a long-run demand elasticity for tinplate of – 0.27 for the US and – 0.21 for Western Europe. Of course, this is hardly reassuring for tin producers. It implies that a higher market share resulting from lower prices will lead to a fall in total sales revenue from tin.

13. Although *TI* (December 1992) noted that the fall in the tin price in Malaysian Ringgit in late 1992 in part could be blamed on the strengthening of the Ringgit against the US dollar, overall in the period after the tin collapse the index of the nominal effective exchange rate of the Ringgit (i.e the trade-weighted exchange rate index, without adjustment for Malaysian inflation *vis-à-vis* the rest of the world) fell substantially. It had reached 70.3 in 1992 (1985=100), though this is a slight rise compared to 1991 and 1990 (IMF, *IFS*, September 1993). This depreciation would have tended to cushion the Ringgit tin price fall over the 1985 to 1992 period. See Figure 6.1 for an illustration of the different behaviour of the tin price in different currencies, primarily as a result of exchange rate changes.

14. LME prices have been published in US dollars since the Exchange reopened in mid-1989.

15. A neat tabular presentation of tin-producing countries' exchange rate changes after the 1985 tin collapse is given in Crowson (1987). According the IMF, *IFS*, (September 1993), the index of Bolivia's real effective exchange rate (the nominal effective exchange rate (see note 13), adjusted for Bolivia's inflation in relation to the rest of the world), with 1985=100, was 29.27 in 1986 and down to 22.04 in 1992, a very large and sustained real depreciation indeed. In contrast, Malaysia's in 1986 was 83.9, and 70.3 in 1992. Unfortunately, the IFS do not give REER figures for Indonesia or Thailand.

16. See Chapter 3, especially note 12.

17. For studies of the price elasticity of supply of tin see Lim (1969), who gives an elasticity based on annual data of 0.28 for the Malaysian gravel pumping sector, though Lim's estimate for Malaysian dredging is not statistically significantly different from zero. A later study, which takes fuller account of supply lags in tin-mining, is Bird (1978). Bird found that supply responses were significant after two years in all major producers, but the full effects took seven years. *A priori* one would expect that the supply response of gravel pumping would be greater than that of dredging because changes in the tin price have a greater effect on gravel pumping's profitability than dredging's (Thoburn, 1977a, 1978a).

The Real Price of Tin and the Terms
of Trade for Tin Producers

In Chapter 1 the real price of tin over the period from the first world war to the present was represented in Figure 1.1 by the New York tin price deflated by the American wholesale price (producer price) index. Before 1914, when Britain rather than the United States was the world's largest consumer of tin, this was substituted by the Sterling tin price deflated by the UK wholesale price index. The sources for these data were shown on Figure 1.1.

While this measure allows for the rate of inflation facing consumers over time, and thus indicates the 'real' price facing consumers, from the producers's stance the purchasing power of a unit of tin exported depends on the price of the imports which can be bought with the foreign exchange earned from the tin. In other words, we need to know the *terms of trade* facing tin producers: P_x/P_m, the price of (tin) exports divided by the average price of imports by tin producers, all measured in a common currency;[1] this is the 'real price' of tin from the producers' viewpoint. However, even assuming that the recipients of factor incomes from tin in, say, Malaysia, have the same import pattern as Malaysians in general, so that we could use Malaysian import unit values to calculate P_m, there is still no easily available data on the average import unit values of tin-producing countries aggregated as a group. Instead, we proxy P_m by taking a price index of the goods they are likely to import: the unit value of manufactures exported by developed market economies,[2] measured, like the tin price used here, in US dollars. These data are published in the UN *International Trade Statistics Yearbook*.[3]

In periods where exchange rates are stable, one would expect American dollar domestic prices to change in a similar fashion to that of the dollar unit values of US exports. There may be some divergence between dollar

FIGURE A.1 Real tin prices and tin terms of trade, 1948–1990 (1960=100)

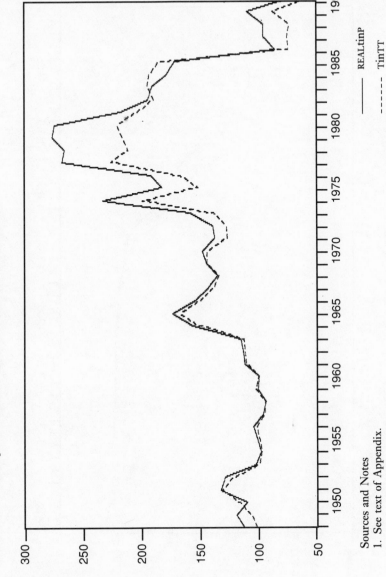

Sources and Notes
1. See text of Appendix.

FIGURE A.2   Behaviour of US producer price index and dollar manufactures price index, 1948–90 (1960=100)

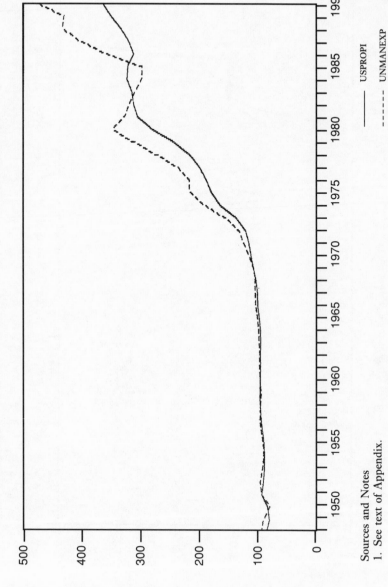

Sources and Notes
1. See text of Appendix.

prices for US exports and DMEs' dollar export prices as a whole, but international competition would help keep American and other DME dollar export prices in line, and US exports in any case are a component of total DME manufactured exports.[4] However, whereas the DMEs' dollar export price index includes, by definition, only traded goods, the US producer price index includes some prices of goods which are not traded. In periods where there are large changes in exchange rates, such as the late 1970s and (especially) the 1980s, series for the two deflators may diverge.[5]

Figure A.1 tracks the behaviour of the real tin price for consumers ('REALTINP') against the real tin price (i.e. the terms of trade) for tin producers ('TINTT'), arbitrarily taking 1960 as a base period, when exchange rates were relatively stable under the Bretton Woods system. Over the 1950s and 1960s the behaviour of the two series is very similar. Over the 1970s the real tin price for consumers rises faster than the tin terms of trade. This reflects the fact that the UN's index of dollar export prices (UNMANEXP) rises in the 1970s in relation to American wholesale prices (see Figure A.2). This relative rise reflects in turn a gradual depreciation of the nominal effective exchange rate of the US dollar over the 1970s, whose index fell to 89.8 in 1973 (1970=100) and, after a slight rise, was down to 78.6 in 1978, and continued to fall slightly to 1980.[6] In consequence of the fact that dollar export prices in the world economy rose faster than US domestic prices, the rise in the tin price in the 1970s to producers in real terms, though substantial, was less than the real price rose to consumers, as Figure A.1 shows.

In the 1980s, as is well-known, the dollar appreciated sharply in nominal (and real) terms up to 1985, and, following the Plaza agreement, depreciated for the rest of the decade. By 1990 it was substantially lower in both nominal and real terms than in the late 1970s. The dollar appreciation explains the dip in UNMANEXP in the early 1980s. As the appreciation peaked in the three years up to 1985, the dip in UNMANEXP relative to USPRODPI caused the tin terms of trade to rise above the real price to consumers. The fall in the dollar in the late 1980s, as in the 1970s, caused the real tin price to consumers to rise relative to that for producers.

Finally, note that we are only discussing here the behaviour of the tin price in relation to the behaviour of the deflators. In other words, the dollar tin price itself is taken as given. Exchange rate changes affect the tin price itself too, of course. As Chapter 6 has shown (see Figure 6.1), in the early 1980s, as the US dollar was appreciating, the tin price fell in dollars (and in Malaysian Ringgit, which closely tracks the US dollar), while rising in £ sterling. Most commodity prices are denominated in US dollars. Generally, dollar appreciation tends to lower dollar commodity prices and dollar depreciation tends to raise them, though the relation is complex.[7] In the early 1980s, when the US dollar appreciated some 40 per cent up to late 1985, dollar

commodity prices fell 25–30 per cent, but the subsequent dollar depreciation did little to raise them (Maizels, 1992, pp. 17–18), possibly because of
the additional supply caused by the debt and balance of payments problems
of LDC producers.

NOTES

1. This is the *net barter* terms of trade, the measure most commonly
   used. One might also consider the *income* terms of trade
   $((P_x.Q_x)/P_m)$, which is the net barter terms of trade weighted by
   export quantity, and which indicates the purchasing power of
   exporting. There is also the *single factoral* terms of trade
   $((P_x.Z_x)/P_m)$, where Z_x is an export productivity index, and some
   other measures. For a full discussion of terms of trade concepts,
   see Meier (1968, pp. 41–6).
2. It would also be possible to take import unit values for imports
   into LDCs as a whole, but, because of differences in import patterns
   between countries, this would not necessarily be more accurate.
   Using the export prices of developed market economies as a deflator is also the method used by UNCTAD to derive 'real prices' (see,
   for example, UNCTAD, 1992a).
3. A series of dollar unit value indices for developed market
   economies' manufactures exports from 1970 to 1990 is published in
   UN (1990b). To take the series back to 1948, I have spliced onto it
   a series, from earlier editions of the Yearbook, of dollar unit value
   indices of *all* manufactured exports. Since, even in 1972, LDCs
   accounted for only 6.6 per cent of world manufactured exports
   (Thoburn, 1977b, p. 4), this series closely approximates to export
   unit values for developed economies' manufactures.
4. In 1989 the United States generated 14.7 per cent of DMEs' exports
   of manufactures (UNCTAD, 1991).
5. For instance, if a country's exchange rate depreciates, this tends to
   raise the domestic currency price of its exports and imports (i.e. of
   its 'traded goods') in relation to the price of domestic goods (such
   as construction) which are not traded. See (e.g.) Dornbusch and
   Helmers (1988) for further analysis. However, the US wholesale
   price index is heavily weighted towards traded goods (and certainly
   more so than is the consumer price index), so the divergence
   between the two deflators used here cannot simply be seen as a
   change in the traded/nontraded goods price ratio. In a large economy such as the USA, where the ratio of exports to national income
   is quite low (about 10 per cent), changes in international prices
   may well not quickly feed through into domestic price changes.
6. Figures on exchange rates are from the IMF's *IFS*, and its *IFS
   Yearbook*.
7. For example, one cannot estimate the effect of a change in the
   dollar exchange rate on each single commodity price in turn, holding prices of other commodities constant; the final result will
   depend on cross supply and demand elasticities for all commodities
   in all trading countries. See Gilbert (1989), who also argues that
   the supply of commodities has been affected by LDCs' debt problems. These problems have forced them into additional supply to

seek export revenue to pay debt service, although the effects are stronger for agriculural products than for minerals. See also Maizels (1992, ch. 1).

Bibliography

Abdullah bin Yusof and Chan Wan Choon (1991), 'The mining history of the Kuala Langat tin deposit as experienced by Petaling Dredging Bhd and Selangor Dredging Bhd', Kuala Lumpur: mimeo (draft of paper for UK Institute of Mining and Metallurgy conference, November 1991).

Abdullah Hasbi bin Haji Hassan and G. R. Wallwork (1984), *Mining Techniques for Alluvial Tin Deposits*, Ipoh: Proceedings of the fourth SEATRAD (South East Asian Tin Research and Development Centre) seminar, 8–11 October.

Allen, G. C., and A. G. Donnithorne (1954), *Western Enterprise in Indonesia and Malaya. A Study in Economic Development*, London: Allen and Unwin.

Anderson, R. W., and C. L. Gilbert (1988), 'Commodity agreements and commodity markets: lessons from tin', *Economic Journal*, March.

ATPC (1983), *A Milestone in Producer Collaboration in Tin*, Kuala Lumpur.

—— *Annual Report*, from 1983–4, Kuala Lumpur.

Auty, R. M. (1987), 'Producer homogeneity, heightened uncertainty and mineral market rigidity', *Resources Policy*, September.

—— (1991a), 'Managing mineral dependence: Papua New Guinea 1972–89', *Natural Resources Forum*, May.

—— (1991b), 'Mismanaging mineral dependence: Zambia 1970–90', *Resources Policy*, September.

Ayub, M. A., and H. Hashimoto (1985), *The Economics of Tin Mining in Bolivia*, Washington: World Bank.

Baldwin, R. E. (1963), 'Export Technology and Development from a Subsistence Level', *Economic Journal*, March.

Baldwin, W. L. (1983), *The World Tin Market: Political Pricing and Economic Competition*, Durham, N.C.: Duke University Press.

Barnett, H. J. (1979), 'Scarcity and growth revisited', in V. K. Smith (ed.), *Scarcity and Growth Reconsidered*, Baltimore: John Hopkins Press, for Resources for the Future.

—— and C. Morse (1963), *Scarcity and Growth: the Economics of Natural Resource Availability*, Baltimore: John Hopkins Press for Resources for the Future.

Bird, P. J. W. N. (1978), 'The price elasticity of supply of tin, 1961–75', *Malayan Economic Review*, October.

Bomsell, O., with I. Marques, D. Ndiaye, and P. deSa (1990), *Mining and Metallurgy Investment in the Third World: the End of Large Projects?*, Paris: Organisation for Economic Cooperation and Development.

Bosson, R., and B. Varon (1977), *The Mining Industry and the Developing Countries*, London: Oxford University Press.

Brodsky, D. A., and G. P. Sampson (1980), 'Retained value and export performance of developing countries', *Journal of Development Studies*, October.

Bureau of Mines (1987), *An Appraisal of Minerals Availability for 34 Commodities: Tin and Tantalum*, Denver: United States Department of the Interior, Bulletin 692.

Burke, G. (1990), 'The rise and fall of the international tin agreements', in K. S. Jomo (ed.), *Undermining Tin: the Decline of Malaysian Preeminence*, Sydney: University of Sydney Transnational Corporations Research Project.

Chai Hon Chan (1964), *The Development of British Malaya, 1896–1909*, Kuala Lumpur: Oxford University Press.

Chulalongkorn University (1976) 'Report of research on small boat mining', Bangkok: Mining Engineering Faculty, mimeographed (in Thai).

CIS (1972), *The Rio-Tinto Zinc Corporation Ltd Anti-Report*, London: Counter Information Services.

Coote, B. (1992), *The Trade Trap: Poverty and the Global Commodity Markets*, Oxford: Oxfam.

Crabtree, J., G. Duffy, and J. Pearce (1987), *The Great Tin Crash: Bolivia and the World Market*, London: Latin American Bureau.

Crowson, P. (1987), 'Tin: the implications of present prices for mines and smelters', *Tin International*, February.

———— (1991), 'Foreign investment in natural resources: a one-way pendulum?', *Institute of Development Studies Bulletin*, Sussex, April.

———— (1992a), 'Geographical shifts in the competitive strength of mineral production since 1960, and their causes', *Resources Policy*, December.

———— (1992b), *Minerals Handbook 1992–3: Statistics and Analyses of the World's Minerals Industry*, London: Macmillan.

Cushman, J. W. (1991), *Family and State: the Formation of a Sino–Thai Tin-Mining Dynasty, 1797–1932*, Singapore: Oxford University Press.

Daniel, P. (1990), 'Economic policy in mineral exporting countries: what have we learned?', Brighton: University of Sussex Institute of Development Studies Discussion Paper No.279, November

DMR (1976), *Mineral Statistics of Thailand, 1971–76*, Bangkok.

———— (1990), *Mineral Statistics of Thailand, 1986–1990*, Bangkok.

DNPM (1990), *Anuario Mineral Brasileiro 1990*, Brasilia.

Dornbusch, R., and F. L. C. H. Helmers (1988), *The Open Economy: Tools for Policy Makers in Developing Countries*, Washington, DC: Oxford University Press for the World Bank.

Dunkerley, J. (1984), 'Bolivia in focus', *Tin International*, September.

Dunning, J., and J. Cantwell (1987), *Institute for Research and Information on Multinationals – Directory of Statistics of International Investment and Production*, London: Macmillan.

EIU (1991), *World Commodity Outlook 1992: Industrial Raw Materials*, London.

Elliott, T. (1989), 'The "Restructuring" of the Malaysian Tin Industry under the New Economic Policy', University of London MSc thesis.

Engel, B. C. (1980), 'Report on visit to Brazil, 8–20 December 1979', London: International Tin Council, mimeo.

———— and H. W. Allen (1979), *Tin Production and Investment*, London: International Tin Council.

Ericsson, M. (1991), 'Minerals and metals production technology: a survey of recent developments', *Resources Policy*, December.

ESCAP (1984), *Mining Taxation in the ESCAP Region: Review and Proposals for Reform*, Bangkok: United Nations.

———— (1985), *Transnational Corporations and Primary Commodity Exports from Selected Developing Countries*, Bangkok: United Nations.

Falkus, M. (1989), 'Early British Business in Thailand', in R. P. T. Davenport-Hines and G. Jones (eds), *British Business in Asia since 1860*, Cambridge: Cambridge University Press.

Farzin, Y. H. (1992), 'The time path of scarcity rent in the theory of exhaustible resources', *Economic Journal*, July.

Fermor, Sir L. L. (1939), *Report upon the Mining Industry of Malaya*, Kuala Lumpur: Government Printer.

Findlay, C., and R. Garnaut (eds) (1986), *The Political Economy of Manufacturing Protection: Experiences of ASEAN and Australia*, London: Allen and Unwin.

Fox, D. J. (1970), *Tin and the Bolivian Economy*, London: Latin American Publications Fund.

Fox, W. (1974), *Tin: the Working of a Commodity Agreement*, London: Mining Journal Books.

Frame, P. K. (1992), 'The year 2001 ... will there be enough tin?', in International Tin Research Institute, *Fifth International Tinplate Conference, 1992, Proceedings*, London.

Freund, B. (1981), *Capital and Labour in the Nigerian Tin Mines*, New Jersey: Humanities Press.

Fryer, D. W., and J. C. Jackson (1977), *Indonesia*, London: Ernest Benn.

Garnaut, R., and A. Clunies Ross (1983), *Taxation of Mineral Rents*, Oxford: Clarendon Press.

Geddes, C. F. (1972), *Patino, the Tin King*, London: Robert Hale.

Gilbert, C. L. (1987), 'International commodity agreements: design and performance', *World Development*, May.

——— (1989), 'The impact of exchange rates and developing country debt on commodity prices', *Economic Journal*, September.

Gillis, M., and R. E. Beals (1980), *Tax and Investment Policy for Hard Minerals: Public and Multinational Enterprises in Indonesia*, Cambridge, Massachusetts: Ballinger .

——— with M. W. Bucovetsky, G. P. Jenkins, U. Peterson, L. T. Wells, and B. D. Wright (1978), *Taxation and Mining: Nonfuel Minerals in Bolivia and other Countries*, Cambridge, Massachusetts: Ballinger.

Girvan, N. P. (1987), 'Transnational corporations and non-fuel primary commodities in developing countries', *World Development*, May.

Hallwood, C. P. (1979), 'The profitability of the buffer stocks operated under the International Tin Agreements, 1956–77', *Resources Policy*, December .

Harvey, C. (1991), 'Come back equity; all is forgiven', *Institute of Development Studies Bulletin*, Sussex, April.

Hedges, E. S. (1964), *Tin in Social and Economic History*, London: Edward Arnold .

Heidhues, M. F. S. (1991), 'Company island: a note on the history of Belitung', *Indonesia*, April.

——— (1992), *Bangka Tin and Mentok Pepper: Chinese Settlement on an Indonesian Island*, Singapore: Institute of South-East Asian Studies.

Hennart, J.-F. (1986a), 'Internalization in practice: early foreign investments in Malaysian tin mining', *Journal of International Business*, summer.

——— (1986b), 'The tin industry', in M. Casson (ed.), *Multinationals in World Trade: Vertical Integration and the Division of Labour in World Industries*, London: Allen and Unwin.

Hewitt, T. (1992), 'Brazilian industrialization', in T. Hewitt, H. Johnson, and D. Wield (eds), *Industrialization and Development*, Milton Keynes: Open University Press.

Hillman, J. (1984), 'The emergence of the tin industry in Bolivia', *Journal of Latin American Studies*, 2.

—— (1988a), 'Bolivia and the international tin cartel, 1931–41', *Journal of Latin American Studies*, May.

—— (1988b), 'Malaya and the international tin cartel', *Modern Asian Studies*, 2.

—— (1990a), 'Bolivia and British Tin Policy, 1939–45', *Journal of Latin American Studies*, May.

—— (1990b), 'The freerider and the cartel: Siam and the international tin restriction agreements, 1931–41', *Modern Asian Studies*, 2.

Hirschman, A. O. (1958), *The Strategy of Economic Development*, New Haven: Yale University Press.

—— (1977), 'A generalised linkage approach to development, with special reference to staples', *Economic Development and Cultural Change*, Supplement .

HoC (1985–6, 1986–7), *The Tin Crisis*, report from the Trade and Industry Committee, in two volumes, London.

Hosking, K. F. G. (1988), 'The world's major types of tin deposits' in C. S. Hutchinson (ed.), *Geology of Tin Deposits in Asia and the Pacific*, Berlin: Springer-Verlag, for United Nations Economic and Social Commission for Asia and the Pacific.

Hotelling, H. (1931), 'The Economics of Exhaustible Resources', *Journal of Political Economy*, April.

Hughes, H., and S. Singh (1978), 'Economic rent: incidence in selected metals and minerals', *Resources Policy*, June .

Humphreys, D. (1982), 'A mineral commodity lifecycle? Relationships between production, price and economic resources', *Resources Policy*, September .

IMF (various issues), *International Financial Statistics*, Washington, DC

—— (various issues), *International Financial Statistics Yearbook* , Washington, DC.

Ingram, J. C. (1971), *Economic Change in Thailand, 1850–1970*, Stanford: Stanford University Press.

ITC (1960), *Statistical Yearbook 1960*, London.

—— (1974), *Aspects of the Marketing of Tin*, London .

—— (1985), *World Tin Mining: Operations, Exploration and Developments*, London, second edition.

—— (1986), *International Tin Statistics, 1976–86*, London,.

—— (various: a), *Monthly Statistical Bulletin*, London.

—— (various: b), *Quarterly Statistical Bulletin*, London.

ITRDC (1937), *Statistical Yearbook 1937*, The Hague.

ITSG (1949), *Statistical Yearbook 1949*, The Hague.

—— (1950), *Tin, 1949–50: A Review of the World Tin Industry*, The Hague

Jackson, J. C. (1968), *Planters and Speculators: Chinese and European Agricultural Enterprise in Malaya, 1786–1921*, Kuala Lumpur: University of Malaya Press.

—— (1969), 'Mining in 18th century Bangka: the pre-European exploitation of a "tin island"', *Pacific Viewpoint*, September.

Jomo, K. S. (1990), 'Malaysia's tin market corner', in K. S. Jomo (ed.), *Undermining Tin: the Decline of Malaysian Preeminence*, Sydney: University of Sydney Transnational Corporations Research Project.

Jones, W. R. (1925), *Tinfields of the World*, London: Mining Publications.

Jordan, R. and A. Warhurst (1992), 'The Bolivian mining crisis', *Resources Policy*, March.

Kestenbaum, R. (1991), *The Tin Men: a Chronical of Crisis*, London: Metal Bulletin Books.

Klein, H. S. (1965), 'The creation of the Patino tin empire', *Inter-American Economic Affairs*, Autumn.

KLSE (1990), *Annual Companies Handbook*, Kuala Lumpur.

Knorr, K. E. (1945), *Tin under Control*, Stanford: Stanford University Press.

Kumar, R. (1991), 'Taxation for a cyclical industry', *Resources Policy*, June.

Labys, W. C. (1980), *Market Structure, Bargaining Power, and Resource Price Formation*, Lexington, Ma.: Lexington Books/D. C. Heath.

Lam, N. V. (1978), 'Incidence of tin export taxation in West Malaysia', *The Developing Economies*, December .

Lanning, G., and M. Mueller (1979), *Africa Undermined: Mining Companies and the Underdevelopment of Africa*, London: Penguin.

Lardy, N. R. (1992), 'China's foreign trade', *China Quarterly*, September.

Li, D. J. (1955), *British Malaya: An Economic Analysis*, New York: The American Press.

Lim Chong Yah (1967), *Economic Development of Modern Malaya*, Kuala Lumpur: Oxford University Press.

Lim, D. (1969), 'The supply response of tin miners in Malaysia', *Kajian Ekonomi Malaysia*, December.

―――― (1980), 'Industrial processing and location: a study of tin', *World Development*.

Lim Mah Hui (1981), *Ownership and Control of the One Hundred Largest Corporations in Malaysia*, Kuala Lumpur: Oxford University Press.

Lipton, M. (1977) *Why Poor People Stay Poor. A Study of Urban Bias in World Development*, London: Temple Smith.

Lloyd, B., and E. Wheeler (1977), 'Brazil's mineral development: potential and problems', *Resources Policy*, March.

Lo Sum Yee (1972), *The Development Performance of West Malaysia*, Kuala Lumpur: Heinemann.

Loh, F. K. W. (1988), *Beyond the Tin Mines: Coolies, Squatters and New Villagers in the Kinta Valley, Malaysia, c.1880–1980*, Singapore: Oxford University Press.

MacDonald, N. (1991), *Brazil: a Mask called Progress*, Oxford: Oxfam.

Mackie, J. A. C. (1971), 'The Indonesian economy, 1950–63', in B. Glassburner (ed.), *The Economy of Indonesia, Selected Readings*, Ithaca: Cornell University Press.

Maizels, A. (1984), 'A conceptual framework for the analysis of primary commodity markets', *World Development*, January.

―――― (1987), 'Commodities in crisis: an overview of the main issues', *World Development*, May.

―――― (1992), *Commodities in Crisis: the Commodity Crisis of the 1980s and the Political Economy of International Commodity Policies*, Oxford: Clarendon Press

Malaysia (1991), *Yearbook of Statistics, 1990*, Kuala Lumpur: Department of Statistics.

MB (various issues), daily, London.

MBM (various issues), London.

McKern, R. B. (1976), *Multinational Enterprise and Natural Resources*, Sydney: McGraw-Hill .

―――― and Praipol Koomsup (1988) (eds), *Mineral Processing in the Industrialisation of ASEAN and Australia*, Sydney: Allen and Unwin.

Meier, G. M. (1968), *The International Economics of Development*, New York: Harper.

Meyanathan, S. (1988), 'Malaysia', in McKern and Praipol (1988).

Mikesell, R.F. (ed.) (1971), *Foreign Investment in the Petroleum and Mineral Industries. Case Studies in Investor-Host Country Relations*, Baltimore: John Hopkins.

———— and J. W. Whitney (1987), *The World Mining Industry: Investment Stategy and Public Policy*, Boston: Allen and Unwin.

Minchinton, W. E. (1957), *The British Tinplate Industry: a History*, Oxford: Clarendon Press.

Mining Journal (1991), *Mining Annual Review*, London: Mining Journal Books.

Mitchell, B. R., with P. Deane (1962), *Abstract of British Historical Statistics*, Cambridge: Cambridge University Press.

Mollema, J. C. (1922), *De Ontwikkeling van het Eiland Billiton en van de Billiton Maatschappij*, The Hague: Nijhoff.

MT (various issues), Kuala Lumpur: Malaysian Tin Industry (Research and Development) Board, monthly.

Nankani, G. (1979), 'Development problems of mineral exporting countries', World Bank Staff Working Paper No. 354, Washington, D.C.

———— (1980), 'Development problems of nonfuel mineral exporting countries', *Finance and Development*, March .

Nash, J. (1979), *We Eat the Mines and the Mines Eat Us: Dependency and Exploitation in Bolivian Tin Mines*, New York: Columbia University Press.

———— (1992), *I Spent my Life in the Mines: the Story of Juan Rojas, Bolivian Tin Miner*, New York: Columbia University Press.

National Union of Mineworkers (1979), *Bolivia: Report of an NUM Delegation in July 1979*, London.

ODI (1988), 'Commodity prices: investing in decline?', London: ODI Briefing Paper, March

Page, W. (1977), 'One attempt at taking a long term view to assist decision makers in the world tin industry – an interim report', Brighton: Science Policy Research Unit, University of Sussex, mimeo.

Pajon Sinlapajan (1969), 'Tin dredging in Thailand', in International Tin Council, *A Second Technical Conference on Tin*, Bangkok .

Pallister, D., S. Stewart, and I. Lepper (1988), *South Africa Inc.: The Oppenheim Empire*, London: Yale University Press, revised edition.

Paranapanema (1989), *Paranapanema*, Sao Paulo (company publicity brochure).

Paranapanema (1992), 'An overview of the Brazilian tin industry', *Metal Bulletin* international tin conference paper, Phuket, May, mimeo.

Pearson, S. R., and J. Cownie (1974), *Commodity Exports and African Economic Development*, Lexington: Heath.

Praipol Koomsup (1988), 'Thailand' in McKern and Praipol (1988).

Putucheary, J. J. (1960), *Ownership and Control in the Malayan Economy*, Singapore: Eastern Universities Press .

Rachan Kanjan-Vanit, Pow Kham-Ourai and W. Champion (1969), 'Offshore Dredging of Tin Deposits in South Thailand', in International Tin Council, *A Second Technical Conference on Tin*, Bangkok .

Radetzki, M. (1985), *State Mineral Enterprises: an Investigation into their Impact on International Metal Markets*, Washington, D.C.: Resources for the Future .

———— (1992), 'The decline and rise of the multinational corporation in the metal mineral industry', *Resources Policy*, March.

Raw, C. (1977), *Slater Walker. An Investigation of a Financial Phenomenon*, London: Deutsch.

Redzwan Sumun (1985), 'The international tin agreements', First Symposium of the Department of Mines, Kuala Lumpur, 7–11 October, mimeo.

Rees, J. (1990), *Natural Resources: Allocation, Economics and Policy*, London: Routledge, second edition.

Robertson, W. (1982), *Tin: its Production and Marketing*, London: Croom Helm.

Roemer, M. (1979), 'Resource-based industrialisation in developing countries: a survey', *Journal of Development Economics*, June .

Roskill (1990), *The Economics of Tin*, London: Roskill Information Services Ltd.

—— (1992), *Raw Materials Group: Who owns Who in Mining, 1992*, London: Roskill Information Services Ltd

Schmitz, C. J. (1979), *World Non-Ferrous Metal Production and Prices, 1700–1976*, London: Cassell.

Siew Nim Chee (1961), 'Labour and tin mining in Malaya' in T. H. Silcock (ed.), *Readings in Malayan Economics*, Singapore: Eastern Universities Press.

Singer, H. W. (1950), 'The Distribution of Gains between Investing and Borrowing Countries', *American Economic Review, Papers and Proceedings*, May

Slade, M. E. (1989), 'The two pricing systems for non-ferrous metals', *Resources Policy*, September.

Smith, A. (1980–2), 'A history of tin mining', *Tin International*, in seven instalments, February 1980 to September 1982.

Smith, D. N. and L. T. Wells (1975), *Negotiating Third-World Mineral Agreements: Promises as Prologue*, Cambridge, Massachussetts; Ballinger.

Smith, G. W., and G. R. Schink (1976), 'The International Tin Agreement: a reassessment', *Economic Journal*, December .

Soussan, J. (1988), *Primary Resources and Energy in the Third World*, London: Routledge.

Squire, L., and H. van der Tak (1975), *Economic Analysis of Projects*, Baltimore: John Hopkins University Press for World Bank Research Publications.

Sriwichchar Bunyapisit, and Pithaya Boonying (1977), 'Report on Tin Research Study', Industrial Research Section, Department of Business Economics Ministry of Commerce, Bangkok (in Thai).

SSB (1991), *China Statistical Yearbook, 1991*, Beijing.

Sudarsono, B. (1988), 'Indonesia', in McKern and Praipol (1988).

Sutphin, D. M., A. E. Sabin and B. L. Reed (1990), *International Strategic Minerals Inventory Summary Report: Tin*, US Geological Survey Circular 930–1, Denver, Co.

Tambunan, T. (1990), 'Macroeconomic adjustment process in Indonesia during the 1980s', *Development and South–South Cooperation*, December.

Tan Theong Hean (1966), ' A study of the nature and sources of credit in Chinese tin mining in Malaya', Kuala Lumpur: University of Malaya graduation exercise.

ter Braake, A. L. (1944), *Mining in the Netherlands East Indies*, New York: Institute of Pacific Relations.

Thoburn, J. T. (1973a), 'Exports and economic growth in West Malaysia', *Oxford Economic Papers*, March.

—— (1973b), 'Exports and the Malaysian engineering industry: a case study of backward linkage', *Oxford Bulletin of Economics and Statistics*, May.

—— (1977a), 'Commodity prices and appropriate technology: some lessons from tin mining', *Journal of Development Studies*, October.

—— (1977b), *Primary Commodity Exports and Economic Development: Theory, Evidence and a Study of Malaysia*, London: Wiley.

—— (1978a), 'High prices favour small producers', *Tin International*, February.

—— (1978b), 'Malaysia's tin supply problems', *Resources Policy*, March.

—— (1981a), *Multinationals, Mining and Development: a Study of the Tin Industry*, Aldershot: Gower.

—— (1981b), 'Policies for tin exporters', *Resources Policy*, June.

—— (1982a), 'Tin: agreement or cartel?', *Resources Policy*, September.

—— (1982b), 'Transnational corporations and the distribution of gains in the tin industry of South East Asia', in Economic and Social Commission for Asia and the Pacific, *Transnational Corporations and Primary Commodity Exports from Asia and the Pacific*, Bangkok: United Nations, pp. 201–35.

—— and M. Takashima (1993), 'Improving British industrial performance: lessons from Japanese subcontracting', *National Westminster Bank Quarterly Review*, February

Tilton, J. E. (ed.) (1992), *Mineral Wealth and Economic Development*, Washington, D.C.: Resources for the Future.

Timah, P. N. (1972), *Tin In Indonesia*, Jakarta.

Timah, P. T. (1976), *Tin in Indonesia*, Jakarta.

TI (various issues) (*Tin International*), monthly, London/Kuala Lumpur.

UN (1990a), *Terms of Reference of the International Tin Study Group*, New York (TD/TIN.7/15).

—— (b: various years), *International Trade Statistics Yearbook*, New York.

—— (c: various years), *Yearbook of National Accounts Statistics*, New York.

UNCTAD (1990), *The Tin Industry in Latin America: Technological Options and Opportunities for Growth*, Geneva (UNCTAD/COM/3).

—— (1991), *Handbook of International Trade and Development Statistics*, Geneva (TD/STAT.19).

—— (1992a), *Commodity Yearbook 1992*, Geneva (TD/B/C.1/STAT.9)

—— (1992b), *International Tin Statistics*, No. 5, July (TD/TIN/STAT/5).

UNIDO (1979), *World Industry since 1960: Progress and Prospects*, New York .

United States Embassy, Indonesia (1977), *Industrial Outlook: Minerals*, Jakarta.

van Helten, J.-J., and G. Jones (1989), in R. P. T. Davenport-Hines and G. Jones, *British Business in Asia since 1860*, Cambridge: Cambridge University Press.

Vernon, R. (1973), *Sovereignty at Bay, The Multinational Spread of US Enterprises*, London: Penguin.

Wall, D. (1988), 'Comparative advantage in minerals processing', in McKern and Praipol (1988).

Walrond, G. W. and Raj Kumar (1986), *Options for Developing Countries in Mining Development*, London: Macmillan.

Watkins, M. H. (1963), 'A staple theory of economic growth', *Canadian Journal of Economics and Political Science*, May .

Wattenberg, B. J. (1976), *The Statistical History of the United States. From Colonial Times to the Present*, New York: Basic Books.

WBMS (1993), *World Metal Statistics Handbook, 1993*.

Widya, T. (1984), 'Gravel pump tin mining with various stripping methods at P. T. Koba Tin', in Abdullah and Wallwork (1984).

Widyono, B., (1977) 'Transnational corporations in export-orientated primary commodities: a study of relative bargaining positions and the distribution of gains, case study no. 1: The Tin Industry of Bolivia', United Nations: CEPAL/CTC Joint Unit.

Williamson, D. (1984), 'The tin market: will past lessons be learned?', *Tin International*, September .

Williamson, J. (ed.) (1983), *IMF Conditionality*, Washington, D.C.: MIT Press.

Wit Satyarakwit (1971), 'Tin: a comparison of gravel pump and dredging mining', Bangkok: Thammasat University MEcon thesis (in English).

Wong Lin Ken (1965), *The Malayan Tin Industry to 1914, with special reference to the states of Perak, Selangor, Negri Sembilan and Pahang*, Tucson: University of Arizona Press.

World Bank (1980), *Thailand: Towards a Development Strategy of Full Participation*, Washington, D.C..

—— (1989), *Malaysia: Matching Risks and Rewards in a Mixed Economy Program*, Washington, D.C..

—— (1990), *Price Prospects for Major Primary Commodities*, volume 1: *Summary, Energy, and Metals and Minerals*, Washington, D.C.: International Trade Division, International Economics Department.

—— (1992), *World Development Report 1992*, Washington, D.C..

Yip Yat Hoong (1969), *The Development of the Tin Mining Industry of Malaya*, Kuala Lumpur: University of Malaya Press.

Zaalberg, P. H. A. (1970), 'Offshore Tin Dredging in Indonesia', Transactions, Section A, London: Institution of Mining and Metallurgy.

Zondag, C. H. (1966), *The Bolivian Economy, 1952–65. The Revolution and Its Aftermath*, New York: Praeger.

Zorn, R. (1976), *Manual of Palm Oil/Rubber and Tin Producing Companies*, London: privately published.

Zorn and Leigh-Hunt, Messrs (1974), *Manual of Rubber and Tin Producing Companies*, London: privately published

Index